Water Quality and Sediment Transport Issues in Surface Water

AA004915

IAHS Scientific Assembly 2017

Proceedings of the International Association of Hydrological Sciences (PIAHS) Volume 377, 2018

Port Elizabeth, South Africa
10-14 July 2017

Editors:

G. Mahe
K. Heal
A.B. Gupta
H. Aksoy

ISBN: 978-1-5108-6145-9

Printed from e-media with permission by:

Curran Associates, Inc.
57 Morehouse Lane
Red Hook, NY 12571

Some format issues inherent in the e-media version may also appear in this print version.

The copyright of any article is retained by the author(s) and distributed under the Creative Commons Attribution 3.0 license. http://creativecommons.org/licenses/by/3.0/
Unless otherwise stated, associated published material is distributed under the same license. The original work can be found online at:

http://www.proc-iahs.net/377/index.html

Printed by Curran Associates, Inc. (2018)

International Association of Hydrological Sciences
Centre for Ecology and Hydrology
Wallingford, Oxfordshire OX10 8BB
United Kingdom

Phone: 441 491 692 442
Fax: 441 491 692 448

http://iahs.info/

Additional copies of this publication are available from:

Curran Associates, Inc.
57 Morehouse Lane
Red Hook, NY 12571 USA
Phone: 845-758-0400
Fax: 845-758-2633
Email: curran@proceedings.com
Web: www.proceedings.com

Volume 377, 2018

Water quality and sediment transport issues in surface water IAHS Scientific Assembly 2017
10–14 July 2017
Port Elizabeth, South Africa
Editor(s): G. Mahe, K. Heal, A. B. Gupta, and H. Aksoy

Proc. IAHS, 377, 1–1, 2018
https://doi.org/10.5194/piahs-377-1-2018
© Author(s) 2018. This work is distributed under
the Creative Commons Attribution 4.0 License.

Preface: Water quality and sediment transport issues in surface water

Gil Mahe[1], **Kate Heal**[2], **Akhilendra B. Gupta**[3], **and Hafzullah Aksoy**[4]

[1]IRD, HydroSciences Laboratory, Montpellier University, Montpellier, France
[2]The University of Edinburgh, School of GeoSciences, Edinburgh, Scotland, UK
[3]Department of Civil Engineering, MNIT, Jaipur, India
[4]Istanbul Technical University, Civil Engineering Faculty, Istanbul, Turkey

Correspondence: Gil Mahe (gil.mahe@ird.fr)

Published: 16 April 2018

Sediment transport and water quality are modified by human activities all along river courses. If research focuses only on pristine basins and large dams, little is known about the quality of the waters flowing to the sea. Most rivers around the world regulated to some extent by hydraulic infrastructure, even in developing countries. How river management impacts on water quality and sediment transport from the upper basins to coastal areas is not well known in many countries, especially in the developing world, even though this may have strong and longlasting effects on coastal geomorphology and ecosystems. In a time where many people try to explain the coastal recession that is observed on many coastlines, from the sea level rise and thus from "global change", knowledge of the actual sediment transport to the sea could bring new perspectives, as the reduction of riverine sediment transfer certainly contributes to this recession. One of the associated questions is what is the role of the human impact on these processes? At what speed do these changes take place?

This proceeding volume gathers together communications about water quality and sediment transport monitoring and modeling, especially for large river basins, with a focus on the relationships between estuarine river systems and coastal areas in terms of water quality and sediment load. There are also studies presenting diverse methods for estimating the amount of sediment released to the sea and its variability in time. All these papers were presented as oral or poster communications during the IAHS International Conference of Port Elizabeth, South Africa, July 2017.

Proc. IAHS, 377, 3–7, 2018
https://doi.org/10.5194/piahs-377-3-2018
© Author(s) 2018. This work is distributed under
the Creative Commons Attribution 4.0 License.

A multi-approach and multi-scale study on water quantity and quality changes in the Tapajós River basin, Amazon

Rodolfo Luiz Bezerra Nóbrega[1,a], **Gabriele Lamparter**[1], **Harold Hughes**[1], **Alphonce Chenjerayi Guzha**[2], **Ricardo Santos Silva Amorim**[3], **and Gerhard Gerold**[1]

[1]Department of Physical Geography, Faculty of Geoscience and Geography, University of Göttingen, Göttingen, Germany
[2]USDA Forest Service, International Programs, c/o CIFOR, World Agroforestry Center, Nairobi, Kenya
[3]Department of Soil and Agricultural Engineering, Federal University of Mato Grosso, Cuiabá, MT, Brazil
[a]now at: Department of Geography & Environmental Science, University of Reading, Reading, UK

Correspondence: Rodolfo Luiz Bezerra Nóbrega (r.nobrega@reading.ac.uk)

Received: 7 June 2017 – Revised: 6 November 2017 – Accepted: 15 November 2017 – Published: 16 April 2018

Abstract. We analyzed changes in water quantity and quality at different spatial scales within the Tapajós River basin (Amazon) based on experimental fieldwork, hydrological modelling, and statistical time-trend analysis. At a small scale, we compared the river discharge (Q) and suspended-sediment concentrations (SSC) of two adjacent micro-catchments ($< 1\,\mathrm{km}^2$) with similar characteristics but contrasting land uses (forest vs. pasture) using empirical data from field measurements. At an intermediary scale, we simulated the hydrological responses of a sub-basin of the Tapajós (Jamanxim River basin, $37\,400\,\mathrm{km}^2$), using a hydrological model (SWAT) and land-use change scenario in order to quantify the changes in the water balance components due to deforestation. At the Tapajós' River basin scale, we investigated trends in Q, sediments, hydrochemistry, and geochemistry in the river using available data from the HYBAM Observation Service. The results in the micro-catchments showed a higher runoff coefficient in the pasture (0.67) than in the forest catchment (0.28). At this scale, the SSC were also significantly greater during stormflows in the pasture than in the forest catchment. At the Jamanxim watershed scale, the hydrological modelling results showed a 2 % increase in Q and a 5 % reduction of baseflow contribution to total Q after a conversion of 22 % of forest to pasture. In the Tapajós River, however, trend analysis did not show any significant trend in discharge and sediment concentration. However, we found upward trends in dissolved organic carbon and NO_3^- over the last 20 years. Although the magnitude of anthropogenic impact has shown be scale-dependent, we were able to find changes in the Tapajós River basin in streamflow, sediment concentration, and water quality across all studied scales.

1 Introduction

Southern Amazonia was the first region of Brazil's Amazon area to be exposed to intensive conversion to agricultural lands (Fearnside, 2016). The Tapajós River, an important tributary of the Amazon River, lost in this basin ca. 30 % of forest cover (ca. $500\,000\,\mathrm{km}^2$) by 2016, mainly due to the establishment of agro-industrial farms. The forest loss in this river basin is projected to reach approximately 65 % by 2050 (Soares-Filho et al., 2006).

The understanding of small areas is essential to propose solutions to maintain tropical forest services, such as water and nutrient cycling, in the Amazon (Vedovato et al., 2016). These areas can be well assessed by experimental catchment studies. For example, Bleich et al. (2016) studied 10 small pristine streams in the Tapajós River basin and argue that in case measures of conservation of small catchments are not taken, environmental impacts on regional streams in South Amazonia are expected to increase. Impacts at regional scales have been the concern of the scientific community with regards to the role of tropical forests in the global climate systems, especially the effects of the Amazon deforestation in large scales (Ometto et al., 2011). Lima et al. (2014) argue that large-scale deforestation triggers complex non-linear interactions between the atmosphere and biosphere, which may impair important ecosystem services such as water for agriculture and hydroelectric power generation.

Although it has been reported that deforestation leads to changes in the water cycle in this region (Davidson et al., 2012), the effects of forest clearing on the concentrations of suspended and dissolved materials that are usually seen in small streams are difficult to be detected in larger streams and rivers (Thomas et al., 2004). However, the chemistry of the large rivers in the Amazon that remained relatively unaltered until 2000 was compromised because of the upcoming growing of area occupied by pastures (Neill et al., 2001). Additionally, analyses of land-use change impacts that were usually limited to small plots or experimental catchments are now possible to be applied to larger scales, such as river basins, due to recent improvements in data collection, archiving and distribution (Eshleman, 2004). New evidence shows that the conversion of forest to pasture is manifested in systematic changes in the hydro-climatology cycle with increase in river discharge in large catchments in the Amazon (Souza-Filho et al., 2016).

In this study, we examined the impact of the land-use change on the streamflow and water quality of the Tapajós River basin using different spatial scales and approaches. Our objective is to identify what signatures from the land-use change are possible to observe within and across these scales.

2 Area of study

Our study focus on the Tapajós River basin (ca. 500 000 km^2), which is the fifth largest sub-basin of the Amazon River and covers 7 % of the total Amazon basin (Pavanato et al., 2016). This basin includes 7 of the 41 municipalities where Brazilian Environmental authorities concentrate anti-deforestation efforts due to their high incidence of forest clearing (Bragança, 2015). In order to estimate the impacts of scale, we integrated to our study a sub-basin of the Tapajós, the Jamanxim River basin (37 400 km^2), and a pair of micro-catchments ($< 1\,km^2$) with contrasting land uses (forest vs. pasture) located in the municipality of Novo Progresso, in the Brazilian state of Pará (Fig. 1). The climate in this area is humid tropical with a rainy season from November to May and a dry season that extends from June to October. Mean annual precipitation averages 1900 mm.

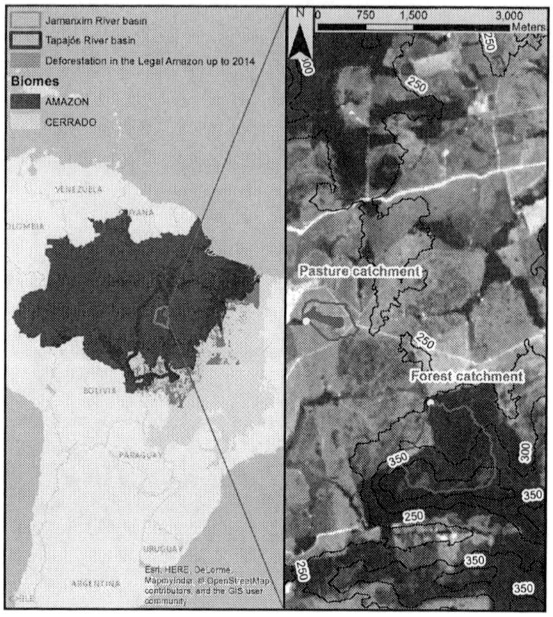

Figure 1. Area of study.

3 Methods

3.1 Experimental micro-catchment study

We compared the streamflow of the micro-catchments by using empirical data from field measurements from 2013 to 2014. At the catchment outlets, we installed rectangular weirs and a DS 5X multiparameter sonde (OTT, USA) to measure water level and to quantify streamflow. We quantified the runoff coefficient as the ratio of total streamflow to total precipitation, and the baseflow index as the ratio of total baseflow to total streamflow following Nóbrega et al. (2017). In these catchments, we also collected 1 L water samples during stormflow events for suspended sediment concentration (SSC) analysis following the method of ASTM (2000). More details on the catchments' characteristics and instrumentation setup can be found in Guzha et al. (2015).

3.2 Jamanxim River basin modelling

We simulated the hydrological behavior of the Jamanxin River basin using the SWAT eco-hydrological model (Arnold et al., 2012). For the setup, calibration and validation of SWAT, we used a gradual land-use change parameterization, field assessments, and available regional data, and then simulated a land-use change scenario in order to quantify the changes in the water balance components due to deforestation. The model parameterization, calibration and validation details can be found in Lamparter et al. (2016). The land-use change scenario used in this study (Fig. 2) suggests a

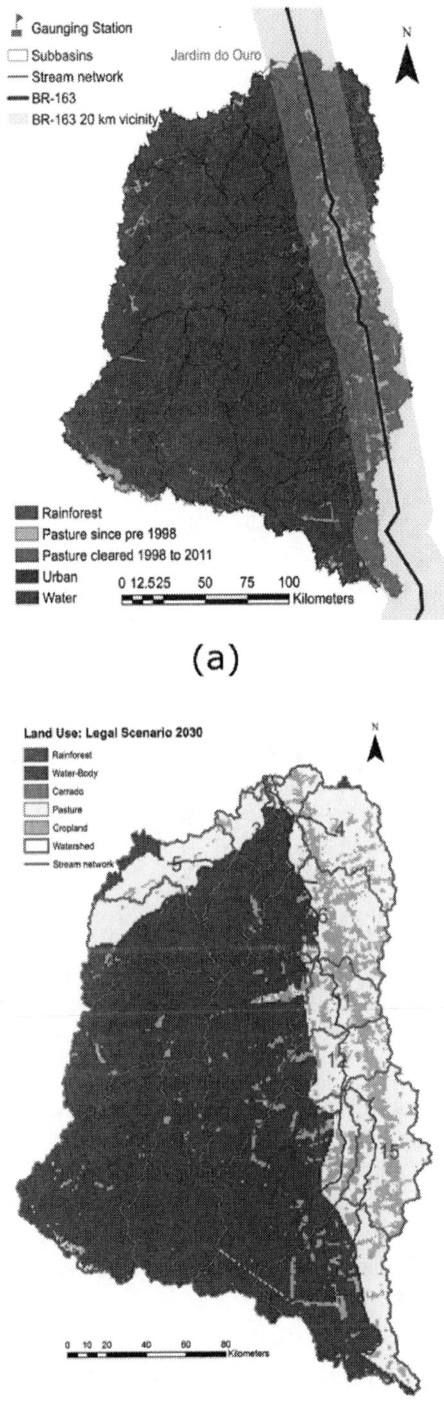

Figure 2. **(a)** Land-use distribution in 2011, and **(b)** Land-use scenario (22 % of deforestation) for the year 2030 following a business as usual approach (Gollow et al., 2017).

Table 1. Q results with SWAT for the land-use distribution and scenario.

P (mm)	Q (mm)	Q scenario (mm)	Qbase (mm)	Qbase scenario (mm)
1639	637	685	396	405

rapid pasture expansion according the study of Gollnow et al. (2017).

3.3 Tapajós River basin analysis

We investigated trends in Q, sediments, and hydrochemistry and geochemistry, i.e. pH, DOC, Mg, K, HCO_3^-, Si, NO_3^- and Ca, in the Tapajós River using available data from the HYBAM Observation Service (http://www.ore-hybam.org, station code 17730000). We used Mann-Kendall test for detecting either an upward or downward trend in the data series with a significance threshold set at .05. The data were also used to quantify fluxes of nitrate and total dissolved carbon (DOC) using mean discharge and concentration in 5-year periods from 1996 to 2015.

4 Results and discussion

Figure 3 shows the streamflow comparison between the two micro-catchments. The pasture catchment has a higher runoff coefficient (0.67) than the forest catchment (0.28). Baseflow indices were 0.76 and 0.88 for the pasture and forest catchments, respectively, showing a higher baseflow contribution in the forest catchment. At this scale, the SSC were also significantly higher during stormflows in the pasture (mean of $579.7\,mg\,L^{-1}$, $n=37$) than in the forest catchment (mean of $81.8\,mg\,L^{-1}$, $n=50$). The geometric mean and 75th percentile for the SSC in the pasture and forest catchments were 47.2 and $26.1\,mg\,L^{-1}$, and 886.0 and $147.8\,mg\,L^{-1}$, respectively.

For the Jamanxim River basin, simulation results show a 2 % increase in discharge (Q) and a 5 % reduction of baseflow contribution to total Q after a 22 % conversion of forest to pasture (Fig. 4 and Table 1). Our results are in accordance to Davidson et al. (2012); they state that even though basin-scale impacts of land use may not yet surpass the magnitude of natural hydrological variability and biogeochemical cycles, there are some signs of a transition to a disturbance-dominated regime, which include changes in the water cycle in the Southern and Eastern regions of the Amazon basin.

At the scale of the Tapajós River basin, however, trend analysis did not show any significant trend in discharge and sediment concentration. Hydrological changes due to land-use change are known to be primarily manifested at smaller scales. Therefore, we ascribe the absence of visible trend at

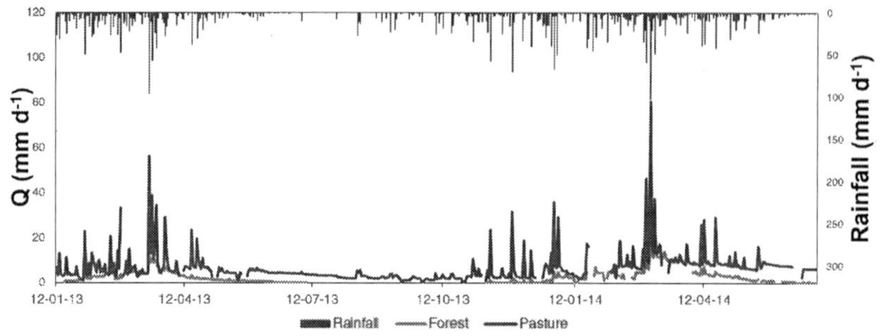

Figure 3. Streamflow and rainfall in the forest and pasture micro-catchments.

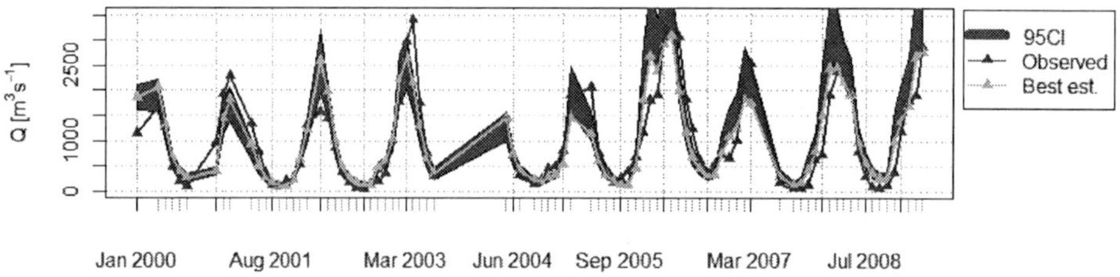

Figure 4. Calibration and validation with land-use update for the Jamanxim catchment.

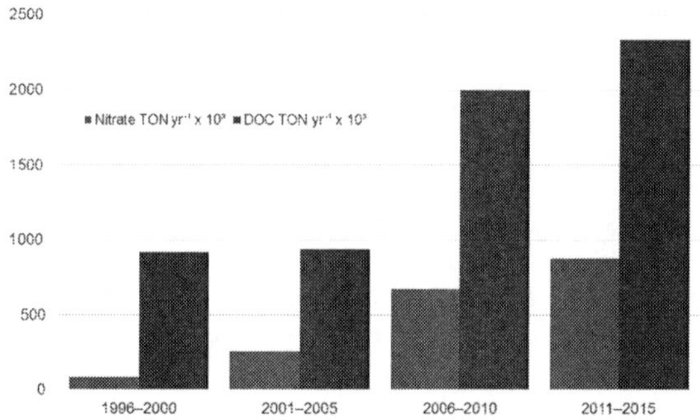

Figure 5. Nitrate and total dissolved carbon fluxes.

a large scale to the fact that most of the deforestation in the Tapajós River basin has occurred in its upper portion, which produces hydrological signatures that may be buffered along the river until its outlet. The analyses of the outflow fluxes over the last 20 years in the Tapajós River revealed upward trends in dissolved organic carbon and NO_3^-, which have reached an up to 10-fold increase (Fig. 5).

5 Conclusions

Effects of deforestation on large rivers of the Amazon basin were relatively unknown due to the low degree of connection between large rivers and land uses in these basins (Neill et al., 2001). We were able to find changes in the Tapajós River basin in river discharge, sediment concentration, and water quality across all studied scales. In this context, our study adds to an increasing body of literature showing that although the magnitude of anthropogenic impact has shown to be scale-dependent, some changes are detectable in both small and large rivers in the Amazon.

Data availability. The data used in this study for the micro-catchments and Jamanxim are available from the Open Science Framework (https://doi.org/10.17605/OSF.IO/UCDE7) and the Spatial Data Infrastructure of the Carbiocial Project (http://gdi.carbiocial.de/). Time series used for the trend analysis are available from the HYBAM Observation Service (http://www.ore-hybam.org), and discharge data used to calibrate and validate the hydrological model are available from the HydroWeb platform of the National Water Agency of Brazil (http://hidroweb.ana.gov.br/, station code: 2650000).

Competing interests. The authors declare that they have no conflict of interest.

Special issue statement. This article is part of the special issue "Water quality and sediment transport issues in surface water". It is a result of the IAHS Scientific Assembly 2017, Port Elizabeth, South Africa, 10–14 July 2017.

Acknowledgements. This research was feasible thanks to the support of the Bundesministerin für Bildung und Forschung (BMBF) through its grant to the CarBioCial project (grant number: 01LL0902A). The authors also acknowledge the data availability of HYBAM Observation Service and the National Water Agency of Brazil (ANA).

Edited by: Akhilendra B. Gupta
Reviewed by: Jagdish Kumar Bassin and one anonymous referee

References

Arnold, J. G., Moriasi, D. N., Gassman, P. W., Abbaspour, K. C., White, M. J., Srinivasan, R., Santhi, C., Harmel, R. D., van Griensven, A., Van Liew, M. W., Kannan, N., and Jha, M. K.: SWAT: Model Use, Calibration, and Validation, T. ASABE, 55, 1491–1508, https://doi.org/10.13031/2013.42256, 2012.

ASTM: Standard Test Methods for Determining Sediment Concentration in Water Samples: D3977-97, West Conshohocken, PA, 2000.

Bleich, M. E., Mortati, A. F., André, T., and Piedade, M. T. F.: Structural Dynamics of Pristine Headwater Streams from Southern Brazilian Amazon, River Res. Appl., 32, 473–482, https://doi.org/10.1002/rra.2875, 2016.

Bragança, A.: Prices, land use and deforestation: Evidence from the Tapajós basin, Rio de Janeiro, available at: http://www.inputbrasil.org (last access: 15 May 2017), 2015.

Davidson, E. A., de Araújo, A. C., Artaxo, P., Balch, J. K., Brown, I. F., Bustamante, M. M., Coe, M. T., DeFries, R. S., Keller, M., Longo, M., Munger, J. W., Schroeder, W., Soares-Filho, B. S., Souza, C. M., and Wofsy, S. C.: The Amazon basin in transition, Nature, 481, 321–328, https://doi.org/10.1038/nature10717, 2012.

Eshleman, K. N.: Hydrological Consequences of Land Use Change: A Review of the State-of-Science, in: Ecosystems and Land Use Change, edited by: Defries, R. S., Asner, G. P., and Houghton, R. A., American Geophysical Union, Washington, D.C., https://doi.org/10.1029/153GM03, 2004.

Fearnside, P. M.: Brazil's Amazonian forest carbon: the key to Southern Amazonia's significance for global climate, Reg. Environ. Chang., https://doi.org/10.1007/s10113-016-1007-2, in press, 2016.

Gollnow, F., Göpel, J., deBarros Viana Hissa, L., Schaldach, R., and Lakes, T.: Scenarios of land-use change in a deforestation corridor in the Brazilian Amazon: combining two scales of analysis, Reg. Environ. Chang., https://doi.org/10.1007/s10113-017-1129-1, in press, 2017.

Guzha, A. C., Nobrega, R. L. B., Kovacs, K., Rebola-Lichtenberg, J., Amorim, R. S. S., and Gerold, G.: Characterizing rainfall-runoff signatures from micro-catchments with contrasting land cover characteristics in southern Amazonia, Hydrol. Process., 29, 508–521, https://doi.org/10.1002/hyp.10161, 2015.

Lamparter, G., Nobrega, R. L. B., Kovacs, K., Amorim, R. S., and Gerold, G.: Modelling hydrological impacts of agricultural expansion in two macro-catchments in Southern Amazonia, Brazil, Reg. Environ. Chang., https://doi.org/10.1007/s10113-016-1015-2, in press, 2016.

Lima, L. S., Coe, M. T., Soares Filho, B. S., Cuadra, S. V., Dias, L. C. P., Costa, M. H., Lima, L. S. and Rodrigues, H. O.: Feedbacks between deforestation, climate, and hydrology in the Southwestern Amazon: Implications for the provision of ecosystem services, Landsc. Ecol., 29, 261–274, https://doi.org/10.1007/s10980-013-9962-1, 2014.

Neill, C., Deegan, L. A., Thomas, S. M., and Cerri, C. C.: Deforestation for pasture alters nitrogen and phosphorus in small Amazonian streams, Ecol. Appl., 11, 1817–1828, https://doi.org/10.1890/1051-0761(2001)011[1817:DFPANA]2.0.CO;2, 2001.

Nóbrega, R. L. B., Guzha, A. C., Torres, G. N., Kovacs, K., Lamparter, G., Amorim, R. S. S., Couto, E., and Gerold, G.: Effects of conversion of native cerrado vegetation to pasture on soil hydro-physical properties, evapotranspiration and streamflow on the Amazonian agricultural frontier, PLoS One, 12, e0179414, https://doi.org/10.1371/journal.pone.0179414, 2017.

Ometto, J. P., Aguiar, A. P. D., and Martinelli, L. A.: Amazon deforestation in Brazil: effects, drivers and challenges, Carbon Manage., 2, 575–585, https://doi.org/10.4155/cmt.11.48, 2011.

Pavanato, H. J., Melo-Santos, G., Lima, D. S., Portocarrero-Aya, M., Paschoalini, M., Mosquera, F., Trujillo, F., Meneses, R., Marmontel, M., and Maretti, C.: Risks of dam construction for South American river dolphins: A case study of the Tapajós River, Endanger. Species Res., 31, 47–60, https://doi.org/10.3354/esr00751, 2016.

Soares-Filho, B. S., Nepstad, D. C., Curran, L. M., Cerqueira, G. C., Garcia, R. A., Ramos, C. A., Voll, E., McDonald, A., Lefebvre, P., and Schlesinger, P.: Modelling conservation in the Amazon basin, Nature, 440, 520–523, https://doi.org/10.1038/nature04389, 2006.

Souza-Filho, P. W. M., de Souza, E. B., Silva Júnior, R. O., Nascimento, W. R., Versiani de Mendonça, B. R., Guimarães, J. T. F., Dall'Agnol, R., and Siqueira, J. O.: Four decades of land-cover, land-use and hydroclimatology changes in the Itacaiúnas River watershed, southeastern Amazon, J. Environ. Manage., 167, 175–184, https://doi.org/10.1016/j.jenvman.2015.11.039, 2016.

Thomas, S. M., Neill, C., Deegan, L. A., Krusche, A. V., Ballester, V. M., and Victoria, R. L.: Influences of land use and stream size on particulate and dissolved materials in a small Amazonian stream network, Biogeochemistry, 68, 135–151, https://doi.org/10.1023/B:BIOG.0000025734.66083.b7, 2004.

Vedovato, L. B., Fonseca, M. G., Arai, E., Anderson, L. O., and Aragão, L. E. O. C.: The extent of 2014 forest fragmentation in the Brazilian Amazon, Reg. Environ. Change, 16, 2485–2490, https://doi.org/10.1007/s10113-016-1067-3, 2016.

Proc. IAHS, 377, 9–18, 2018
https://doi.org/10.5194/piahs-377-9-2018
© Author(s) 2018. This work is distributed under
the Creative Commons Attribution 4.0 License.

Changes in soil erosion and sediment transport based on the RUSLE model in Zhifanggou watershed, China

Lei Wang, Ju Qian, Wen-Yan Qi, Sheng-Shuang Li, and Jian-Long Chen

College of Resources and Environmental Sciences, Lanzhou University, Lanzhou 730000, China

Correspondence: Ju Qian (qianju@lzu.edu.cn)

Received: 8 June 2017 – Revised: 19 January 2018 – Accepted: 6 February 2018 – Published: 16 April 2018

Abstract. In this paper, changes of sediment yield and sediment transport were assessed using the Revised Universal Soil Loss Equation (RUSLE) and Geographical Information Systems (GIS). This model was based on the integrated use of precipitation data, Landsat images in 2000, 2005 and 2010, terrain parameters (slope gradient and slope length) and soil composition in Zhifanggou watershed, Gansu Province, Northwestern China. The obtained results were basically consistent with the measured values. The results showed that the mean modulus of soil erosion is 1224, 1118 and 875 $t\,km^{-2}\,yr^{-1}$ and annual soil loss is 23 130, 21 130 and 16 536 in 2000, 2005 and 2010 respectively. The measured mean erosion modulus were 1581 and 1377 $t\,km^{-2}\,yr^{-1}$, and the measured annual soil loss were 29 872 and 26 022 t in 2000 and 2005. From 2000 to 2010, the amount of soil erosion was reduced yearly. Very low erosion and low erosion dominated the soil loss status in the three periods, and moderate erosion followed. The zones classified as very low erosion were increasing, whereas the zones with low or moderate erosion were decreasing. In 2010, no zones were classified as high or very high soil erosion.

1 Introduction

Soil erosion is a process which refers to the destruction, separation, removal and sedimentation of the earth's surface soil and its parent material caused by hydraulic, wind, freezing and thawing, gravity and other external forces (Meyer, 1984). Soil erosion has caused a set of ecological and environmental problems such as land degradation, soil fertility loss, river siltation, making it a global research focus. Since the 1950s, to quantify soil loss and determine its risk, a number of soil erosion models were established based on measured data or the results of previous studies, and numerous research results were obtained using Geographical Information Systems (GIS) and remote sensing (RS). One of the most applied models to estimate soil erosion is the Universal Soil Loss Equation (USLE) and its modified version the Revised Universal Soil Loss Equation (RUSLE). Lu et al. (2001) used GIS and the RUSLE model to map and quantitatively predict patch and gully erosion in Australia. Liu (2002) established China's Soil Erosion Prediction Equation (CSLE) by the study of a slope erosion prediction model. Gao et al. (2015) calculated the soil erosion modulus in the Loess Plateau of China in 2010 by using the RUSLE model. Because of the complex landscape, the soil erosion status varies in the Loess Plateau. This paper focused on analysing the land use pattern, vegetation distribution and soil erosion status based on the RUSLE model and the impact of slope changes and land use types on erosion by using RS and GIS in the Zhifanggou watershed, Loess Plateau. The aim was to analyse the evolution of soil erosion in a complete and systematic way and provide a reference basis for soil and water loss prevention in hilly and gully regions of the Loess Plateau.

2 Materials and methods

2.1 Study area

Zhifanggou watershed is located in the gully area of Longdong Loess Plateau, Gansu Province, Northwestern China. This watershed has been a key management and experimental area in the Yellow River Basin since 1954. Many management approaches have been carried out to reduce soil erosion, such as soil and water conservation measures, renovating barren slopes and terraces. Ten hydrology and rainfall

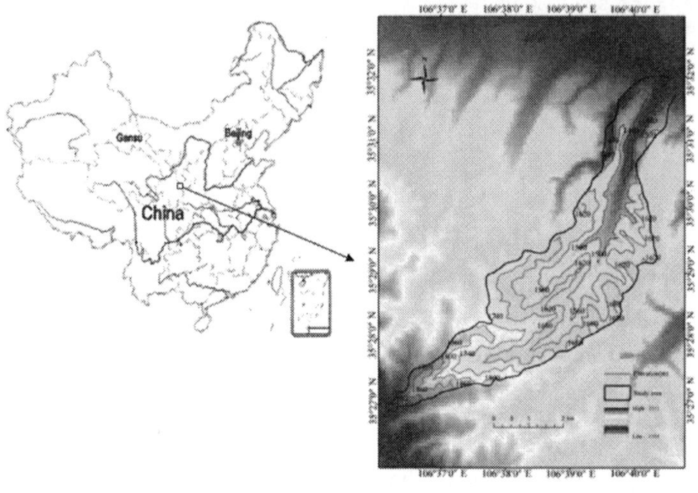

Figure 1. The topography of Zhifanggou watershed.

observation stations were set up to observe rainfall, evaporation, runoff, flood and sediments for more than 50 years.

Figure 1 presents the Zhifanggou watershed. It is one of the tributaries to the Jinghe river and is located south of Pingliang City, China. The area of Zhifanggou watershed is 18.98 km^2. It extends between the latitudes of 35°26′ and 35°33′ N and longitudes of 106°37′ and 106°42′ E. The watershed altitude ranges from 1365 to 2104 m. The main channel length is 15.77 km and the mean channel slope is 4.34 %.

2.2 RUSLE model

The factors in the RUSLE model include rainfall erosivity, soil erodibility, slope length and steepness, vegetation cover and management, and supporting practices (Wischmeier and Smith, 1965). In order to estimate the soil loss, meteorological data, soil information, RS data and a digital elevation model (DEM) were used as detailed below.

1. Three Landsat Thematic Mapper images of the study area from May 2000, May 2005 and July 2010, with 30 m of spatial resolution were used.

2. Daily rainfall data from 1981 to 2004 were obtained from the first and second dams and Hejiazhuang hydrological stations, and Majiaxinzhuang, Chenjiazhuang and Shiyaojian precipitation stations. Because of the lack of observers and funds, the observations were withdrawn in 2005 and merged into one station, named Pingliang hydrological station. Therefore annual rainfall data of Pingliang hydrological station were used from 2005 to 2010.

3. A DEM of 30 m spatial resolution was obtained.

4. Soil information (including organic matter content and texture) for the study area was obtained from the analysis of soil samples.

The RUSLE model is used to estimate soil loss by water-caused erosion (Renard et al., 1997; Duarte et al., 2016). The mathematical expression of RUSLE is

$$A = R \times K \times LS \times C \times P \tag{1}$$

where A is the mean annual soil loss ($\mathrm{t\,km^{-2}\,yr^{-1}}$), caused by rainfall and runoff. R is the rainfall erosivity factor ($\mathrm{MJ\,mm\,hm^{-2}\,h^{-1}}$), which reflects the potential soil erosion caused by rainfall. In RUSLE, it is defined as the product of rainfall energy and maximum rainfall intensity of 30 min. K is the soil erodibility factor ($\mathrm{t\,h\,MJ^{-1}\,mm^{-1}}$), which represents

the amount of soil erosion caused by unit rainfall force in the standard plot. It is used to reflect soil sensitivity to erosion. The slope length and steepness (LS) factor represents the effect of topography on soil erosion, L is the slope length factor and defined as a power function of slope length. S is the slope steepness factor. The vegetation cover factor (C) represents the ratio of soil loss from an area with specified cover and management to soil loss from an identical area in continuous fallow. The conservation support practice factor (P) refers to the ratio of amount of soil loss when soil and water conservation measures, such as contouring or terracing, have been implemented to the original soil loss from sloping land. (Renard et al., 1991; Renard and Ferreira, 1993; Farhan and Nawaiseh, 2015). Soil losses estimated from RUSLE were compared with the measured values which were calculated using the measured annual runoff and suspended sediment of Zhifanggou watershed.

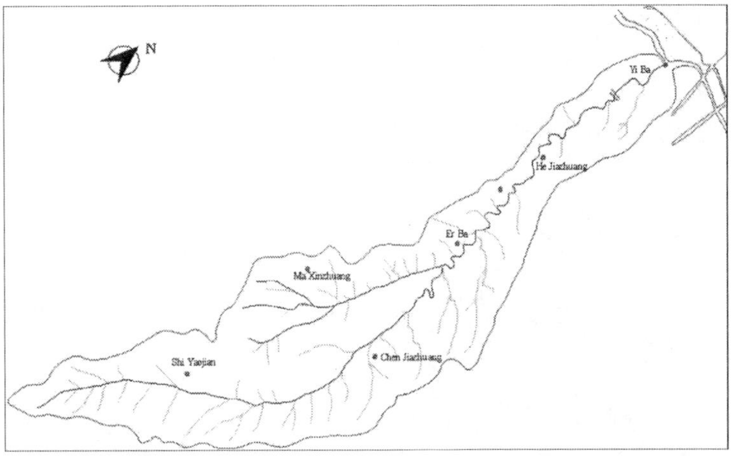

Figure 2. The distribution of rainfall and hydrologic stations.

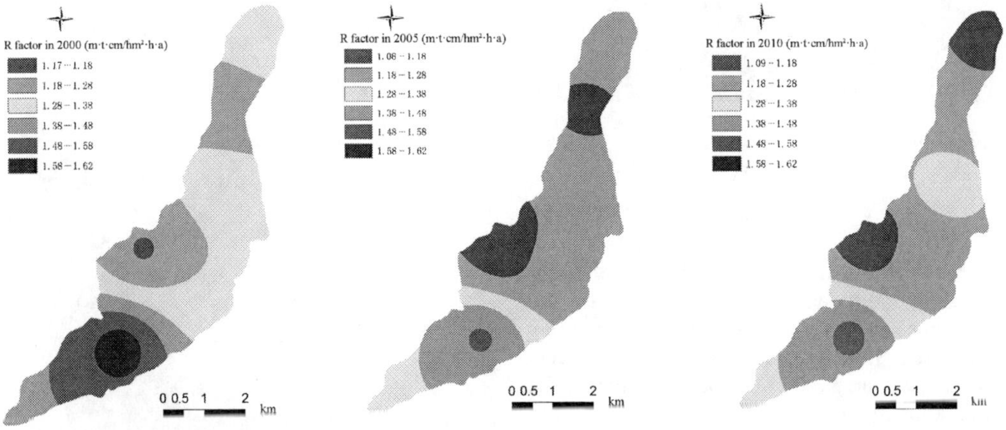

Figure 3. The spatial distribution of rainfall erosivity factor.

2.2.1 Rainfall erosivity factor (*R*)

The *R* factor reflects the potential soil erosion caused by rainfall. In this paper, *R* was calculated from the original algorithm of

the USLE model proposed by Wischmeier and Smith (1978). The equation is:

$$R=\sum E\cdot I_{30} R=\sum E\cdot I_{30} \tag{2}$$

where E is the kinetic energy in a single rainfall event ($\mathrm{MJ\,hm^{-2}}$). It may be counted as a single event if the rainfall is intermittent within 2 h, otherwise the rainfall will be treated twice; I_{30} is the maximum rainfall intensity in 30 min ($\mathrm{mm\,h^{-1}}$).

The equation used to calculate E is:

$$m_{\mathrm{e}}=\begin{cases}0.119+0.0873lg\,(m_{\mathrm{i}}),m_{\mathrm{i}}\leq 76\\ 0.283,m_{\mathrm{i}}>76\end{cases} \tag{3}$$

$$E=\sum m_{\mathrm{e}}\cdot P_m \quad E=\sum m_{\mathrm{e}}\cdot P_m \tag{4}$$

Where m_{e} is the unit kinetic energy during one rainfall event ($\mathrm{MJ\,mm^{-1}\,hm^{-2}}$); m_{i} is the rainfall intensity corresponding to a certain period ($\mathrm{mm\,h^{-1}}$); P_m is the total rainfall corresponding to a certain period (mm). Based on the interpolation and extension of rainfall data, mean annual rainfall erosivity of six stations were calculated during three periods: before 2000, 2001 to 2005 and 2006 to 2010. The distribution of the six stations is shown in Fig. 2. The rainfall erosivity map (Fig. 3) was obtained by the inverse distance weighting method (IDW) which has been shown to be most suitable for rainfall interpolation in this area of China (Bai et al., 2012).

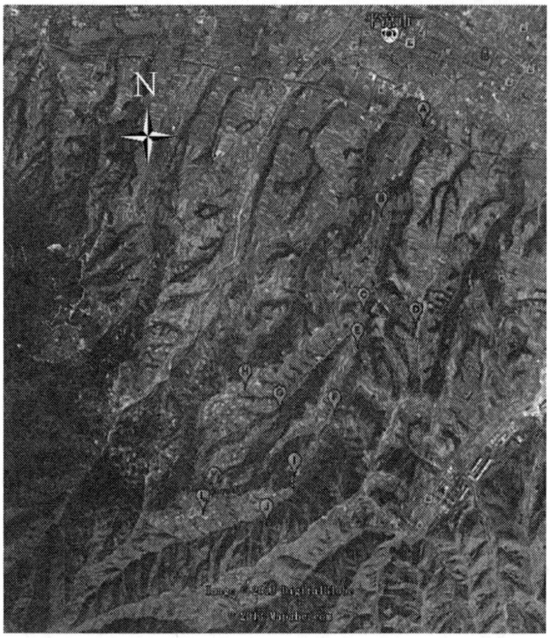

Figure 4. The spatial distribution of soil samples.

2.2.2 Soil erodibility factor (K)

According to the equation, the greater the K value, the greater the risk of soil erosion and vice versa. Twelve soil samples were collected in this watershed and analyzed to determine particle size (sand, silt, clay) and organic matter content in order to calculate K. The distribution of the 12 soil samples is shown in Fig. 4.

There are many methods to calculate the K factor (Wischmeier and Smith, 1978). Considering that the parameters of the EPIC method are measured readily and the method is relatively mature, K was calculated by using the EPIC (erosion-productivity impact calculator) model (Sharpley and Williams, 1990):

$$K = \{0.2 + 0.3\exp[-0.0256\mathrm{SAN}(1.0 - \mathrm{SIL}/100)]\}\left(\frac{\mathrm{SIL}}{\mathrm{CLA}+\mathrm{SIL}}\right)^{0.3} - \left(1.0 - \frac{0.25C}{C + \exp(3.72 - 2.95)}\right)\left(1.0 - \frac{0.7\mathrm{SN1}}{\mathrm{SN1} + \exp(-5.51 + 22.9\mathrm{SN1})}\right) \quad (5)$$

where SAN, SIL, CLA and C are sand, silt, clay and organic carbon (%).

The equation used to calculate SN1 is:

$$\mathrm{SN1} = 1 - \mathrm{SAN}/100. \quad (6)$$

The K factor map (Fig. 5) was obtained by the kriging spatial interpolation method which is more suitable for soil moisture spatial interpolation in the Loess Plateau region (Yao et al., 2013).

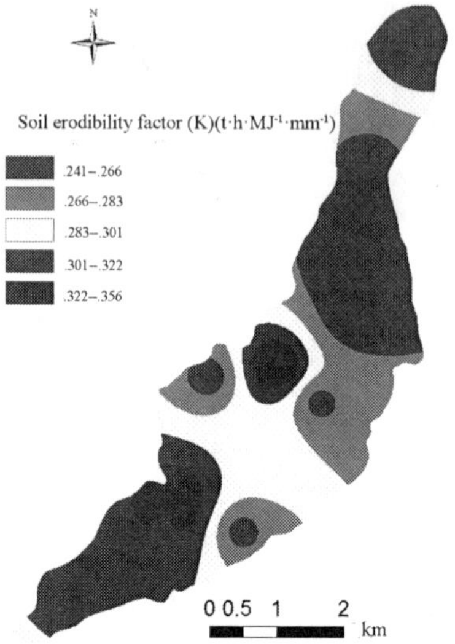

Figure 5. Soil erodibility map (K).

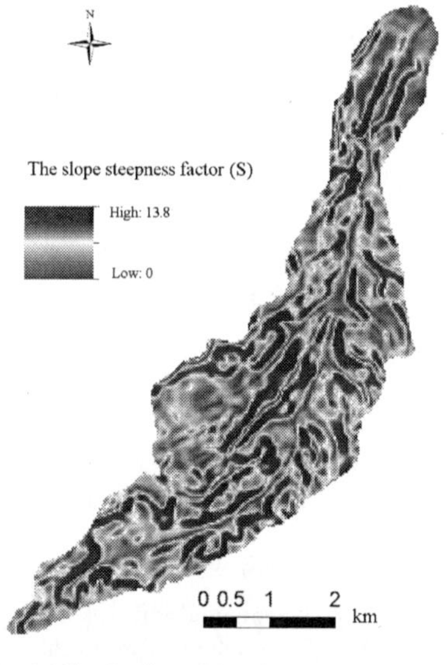

Figure 6. Spatial distribution of the slope steepness factor.

2.2.3 Topographic factor (LS)

The slope steepness factor (S) reflects the influence of slope gradient on erosion (Lu et al., 2001). The greater the mean slope, the greater the potential risk of soil separation and the more severe the soil erosion. S is calculated by using the equations established by Liu et al. (1994) and Liu (2002) as follows:

$$\begin{cases} S = 10.80 \times \sin\theta + 0.03 & \theta < 5^\circ \\ S = 16.80 \times \sin\theta - 0.50 & 5^\circ \le \theta < 10^\circ \\ S = 21.91 \times \sin\theta - 0.96 & \theta \ge 10^\circ \end{cases} \quad (7)$$

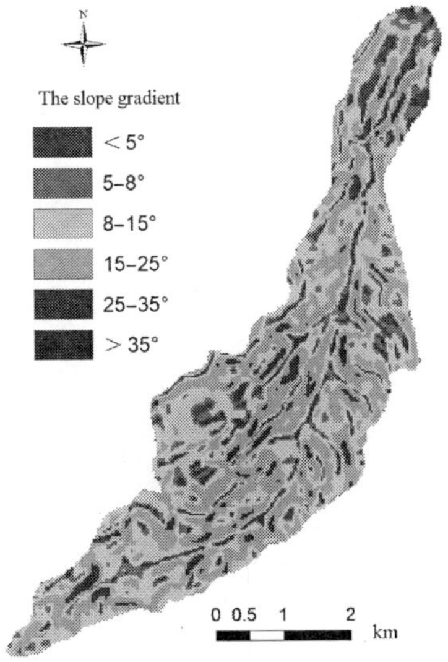

Figure 7. Spatial distribution of slope steepness.

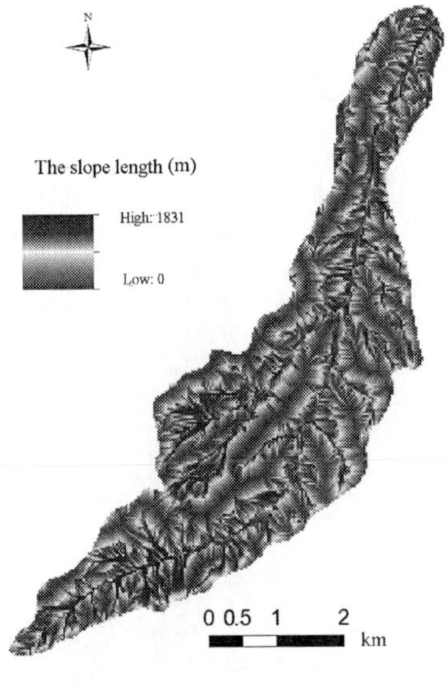

Figure 8. Spatial distribution of slope length.

The DEM data of the basin were used to obtain the gradient grid by using the Surface Analysis function in the Spatial Analyst module of ArcGIS.

The slope data were divided into categories of < 5, 5–8, 8–15, 15–25, 25–35, $\geq 35°$ by using the *Reclassify option in 3D Analyst* module of ArcGIS. The spatial distribution of slope gradient is shown in Fig. 7.

The slope length factor (L) represents the effect of slope length on erosion. The longer the slope length, the greater the flow, and the more severe the soil erosion. L was calculated

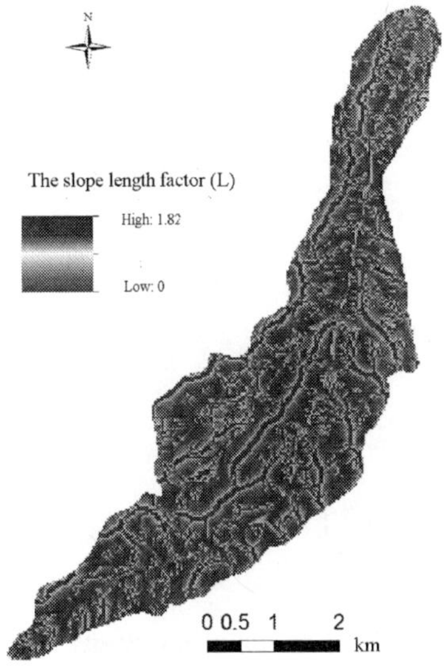

Figure 9. Spatial distribution of L factor values.

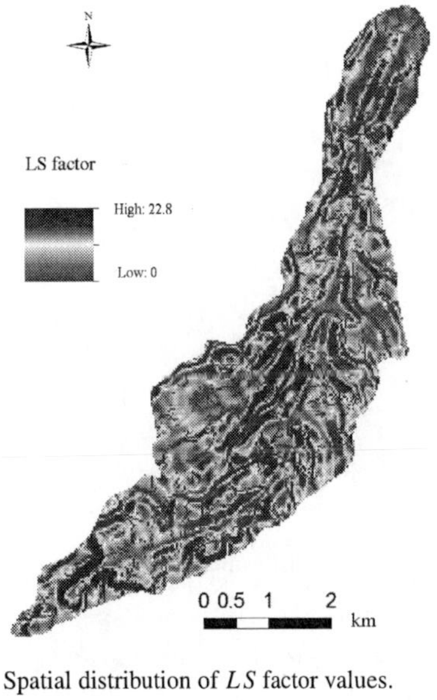

Figure 10. Spatial distribution of LS factor values.

by the classical method of USLE, as follows:

$$L = (\lambda/22.13)^{\alpha} \tag{8}$$

$$\alpha = \beta/(\beta+1) \tag{9}$$

$$\beta = (\sin\theta/0.0896)/\left|3.0\left(\sin\theta^{0.8}+0.56\right)\right| \tag{10}$$

Where λ is the slope length (m), 22.13 represents the slope length of the standard district (m), α is the slope length exponent, β indicates the ratio of rill to interrill erosion and θ represents the slope gradient extracted from the DEM data.

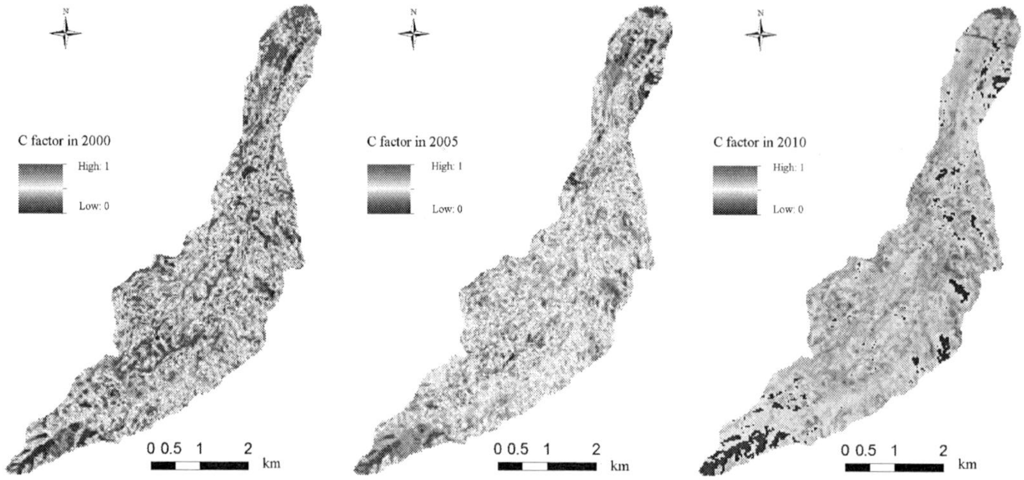

Figure 11. The spatial distribution of vegetation cover factor in the three study years.

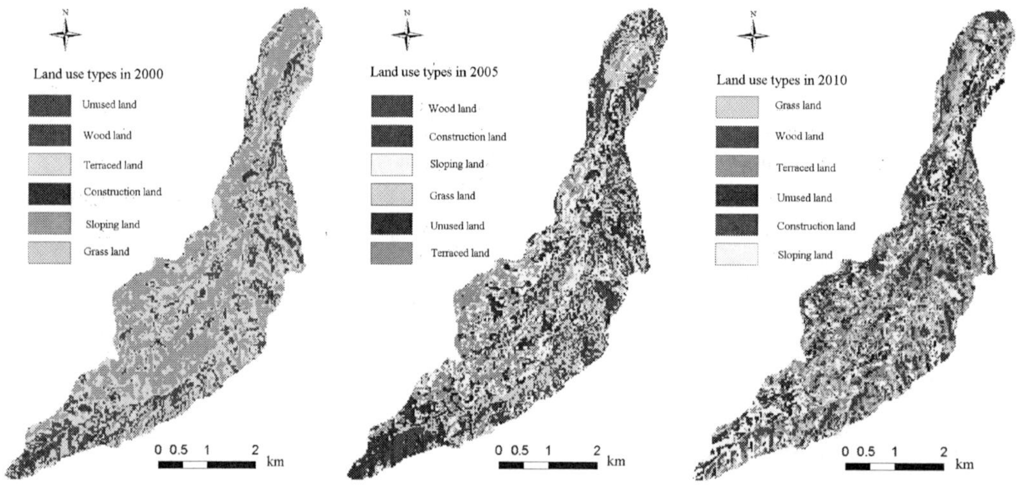

Figure 12. The spatial distribution of land use patterns in the three study years.

The DEM was used to obtain the slope length (Fig. 8) by using ArcGIS.

Considering there is no function to calculate the slope length directly in ArcGIS, we first calculated the flow direction of the grid by the surface runoff simulation. Then we calculated the flow accumulation of the grid. Based on the extraction of the zero value of the flow accumulation, the ridge line was obtained. Finally, the slope length was calculated using the generated ridge line.

Based on the *Map algebra* function of ArcGIS, L was calculated according to the equation and is shown in Fig. 9. Multiplying the spatial distribution of the S and L factors, the LS value distribution is obtained, as shown in Fig. 10.

2.2.4 Vegetation cover factor (C)

The vegetation can effectively protect the surface soil from direct impact by rainfall and play an important role in moderating surface runoff, extending moisture penetration time and enhancing soil impact resistance. The Normalized Difference Vegetation Index (NDVI) reflects the vegetation coverage well. The NDVI data used in this paper (May 2000, May 2005 and July 2010) was collected from the International Scientific Data Service Platform. The vegetation cover is calculated by the classical method (Tan et al., 2005), as follows:

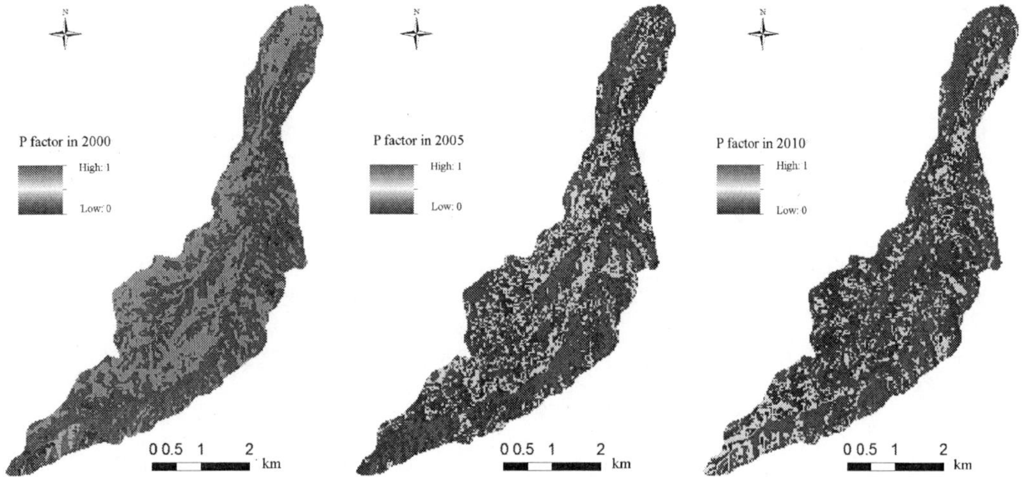

Figure 13. The spatial distribution of the conservation support practice factor in the three study years.

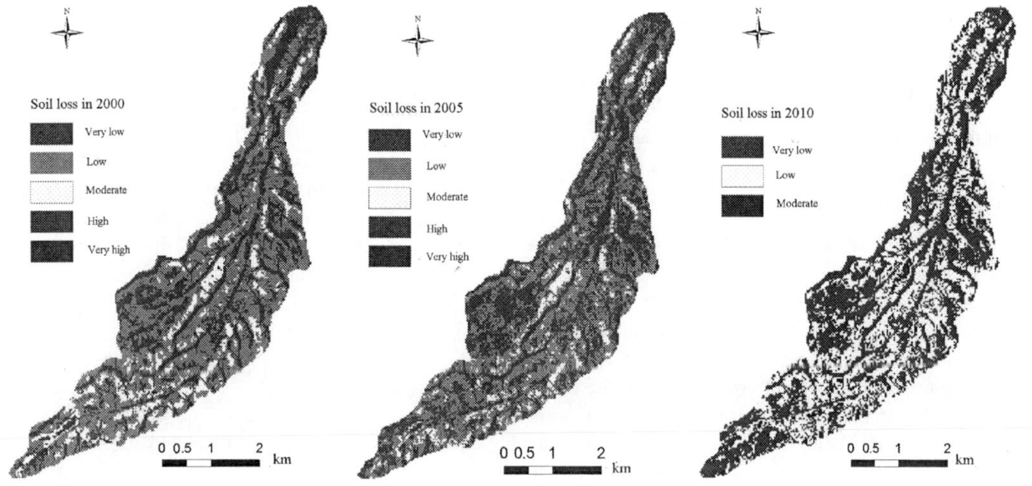

Figure 14. The spatial distribution of soil loss estimated using RUSLE for each of the study years.

$$\mathrm{fg} = (\mathrm{NDVI} - \mathrm{NDVI_{min}})/(\mathrm{NDVI_{max}} - \mathrm{NDVI_{min}}) \times 100 \quad (11)$$

where fg is the vegetation coverage (%), $\mathrm{NDVI_{max}}$ and $\mathrm{NDVI_{min}}$ are the maximum and minimum values of NDVI. By using the method of Cai et al. (2000), C is calculated as follows (in order to prevent a negative C value, the method is slightly modified as below). The spatial distribution of C was calculated for each of the study years 2000, 2005 and 2010 and is shown in Fig. 11.

$$C = \begin{cases} 1 & 0 \le \mathrm{fg} < 0.095 \\ 0.6508 - 0.3436\log_{10}(\mathrm{fg}) & 0.095 \le \mathrm{fg} \le 78.3 \\ 0 & \mathrm{fg} > 78.3 \end{cases} \quad (12)$$

2.2.5 Conservation support practice factor (P)

The value of P ranges from 0 to 1, where 0 indicates good conservation practice and 1 indicates poor conservation practice (Wischmeier and Smith, 1978). The P value was assigned according to Liu et al. (2001), Liu (2006) and Zhao et al. (2016).

RS images in 2000, 2005 and 2010 were classified and reprocessed in ENVI4.8 (Pan et al., 2013) to obtain land use patterns, as shown in Fig. 12. The land use is divided into six categories: terraced land, sloping land, wood land, grass land, construction land and unused land, as shown in Table 1. The P factor was assigned according to field investigation and agricultural practices observed in the study basin. The P val-

Table 1. Conservation support practice factor (P) value.

Land use type	Terraced land	Sloping land	Wood land	Grass land	Construction land	Unused land
P	0.3	0.55	0.9	1	0.25	0.2

Table 2. Areas of different categories of soil loss in the study catchment (ha).

Soil loss categories	The mean erosion modulus standard ($\mathrm{t\,km^{-2}\,yr^{-1}}$)	Soil erosion area by category in 2000 (%)	Soil erosion area by category in 2005 (%)	Soil erosion area by category in 2010 (%)
Very low	< 500	66 747 (35.32)	86 818 (45.94)	104 214 (55.14)
Low	500–2500	97 911 (51.80)	86 791 (45.92)	77 379 (40.94)
Moderate	2500–5000	23 481 (12.42)	15 059 (7.97)	7408 (3.92)
High	5000–8000	834 (0.44)	332 (0.18)	0 (0)
Very high	8000–15 000	28 (0.015)	9 (0.005)	0 (0)

ues of terraced land and construction land were close to 0 due to good soil and water conservation measures. The P values of wood land and grassland were 1, due to the scattered distribution of these land uses and lack of artificial protection measures. The results are assigned to the RS image classification map to obtain raster maps of P for each of the study years, as shown in Fig. 13.

2.2.6 Soil erosion map

The spatial distribution of soil loss in 2000, 2005 and 2010 were obtained from multiplication of the five factors (R, K, LS, C and P) using the grid calculator of *Spatial Analyst*in ArcGIS software.The spatial distribution of the soil loss was obtained and presented in Fig. 14. Table 2 shows the modeled soil loss areas calculated across all grid squares in the basin for each study year according to the classification standards for soil erosion of China in 2007. The total modeled soil erosion amount was calculated as the sum of the modeled soil losses in all grid squares. The mean modelled erosion modulus was calculated by dividing the total soil erosion amount by the total area of the basin. The modeled and measured erosion modulus are shown in Table 3. Because of the lack of observers and funds, soil erosion was not measured in the study area in 2010, so there is no measured value for 2010 in Table 3.

3 Results and discussion

3.1 Estimation of soil erosion

According to Fig. 14 and Table 3, the estimates of soil loss made using RUSLE were basically consistent with the measured values. Soil erosion in the study area is mainly classified as very low and low. In 2000, 2005 and 2010, very low erosion accounted for 35.32, 45.94 and 55.14 % of the total catchment area, low erosion area accounted for 51.80, 45.92 and 40.94 %, and moderate erosion accounted for 12.42, 7.97, 3.92 %. The areas classified as very low erosion are increasing, whilst the areas with low or moderate erosion are decreasing. It can be concluded therefore that the erosion intensity is decreasing. In 2010, no areas were classified as high or very high soil erosion. A similar trend was reported by Gao et al. (2015) of increasing areas of very low erosion in the Loess Plateau from 2010. The parameters in the model can significantly impact the accuracy of the result, so the lack of quantitative description of the P values (conservation support practice factor) may decrease the accuracy of the results. More accurate characterization of P values is an important area for future study.

3.2 Characteristics of soil erosion under different land use patterns

Land use has changed the micro-terrain and original vegetation types and their coverage in the Zhifanggou watershed, affecting the dynamics of soil erosion and resistance. Using the *Spatial analysis* function of ArcGIS, the land use pattern map and the modelled soil loss map in 2010 were superimposed in order to assess the soil erosion characteristics of the different land use types.

Figure 15 shows the intensity of soil erosion is different among different land use types. The soil erosion intensity of terraced fields, sloping land and construction land is mainly very low due to soil and water conservation measures. The erosion intensity of wood land is mainly moderate, and the grassland is mainly low. Because the distribution of wood land is intermixed with natural grassland, and also due to the lack of artificial soil protection measures, the erosion intensity is high. The erosion intensity of unused land is very high, because its surface vegetation is sparse and the land is difficult to manage.

These results are consistent with the findings of Zhang (2016) that the construction of terraced fields reduced

Table 3. Comparison of modeled and measured annual soil loss across the whole study catchment.

Year	The mean erosion modulus $(t\,(km^2\,yr)^{-1})$	Modeled soil loss $(t\,yr^{-1})$	The measured mean erosion modulus $(t\,(km^2\,yr)^{-1})$	Measured soil loss $(t\,yr^{-1})$
2000	1224	23 130	1581	29 872
2005	1118	21 130	1377	26 022
2010	875	16 536		

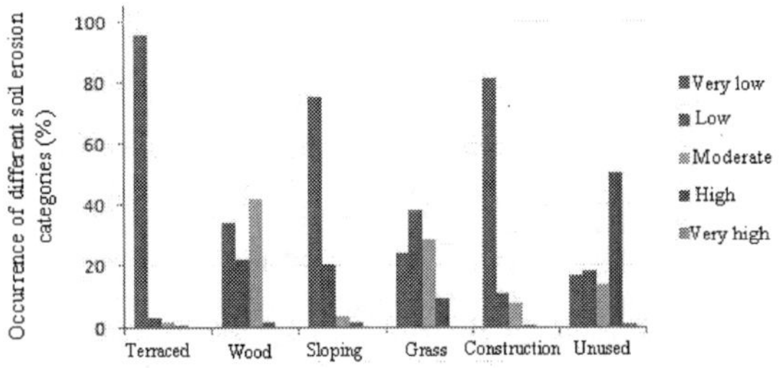

Figure 15. Characteristics of soil erosion under different land use patterns across all three study years.

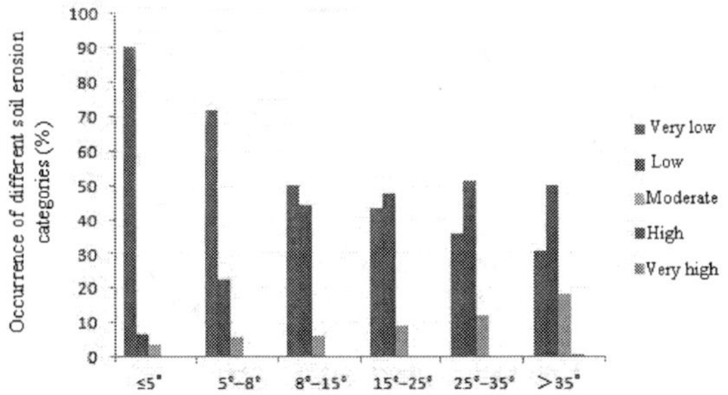

Figure 16. Characteristics of soil erosion under different slope gradient across all three study years.

the area of mild erosion intensity in the basin and increased the area of low erosion intensity, which had a positive effect on soil and water conservation.

3.3 Characteristics of soil erosion under different slope gradients

Using the *Spatial analysis* function of ArcGIS, the slope gradient map and the soil loss map in 2010 were superimposed in order to examine soil erosion intensity under different slope gradients.

Figure 16 shows that the intensity of soil erosion is different among different land slope gradients. The soil erosion intensity of slopes $< 5°$ is mainly very low. Areas of very low erosion decreased and low and moderate erosion increased in the 5–15° slope categories. Areas of very low and low erosion decreased and moderate erosion increased in the 15–35° slope categories. On slopes greater than 35°, there is some high soil erosion and the moderate erosion area increases. In conclusion, in the same conditions, estimated soil erosion increased obviously with the increase of slope in the range of 0–15°. When the slope is greater than 15°, soil erosion intensity increased only slightly with the increase of slope. These results are consistent with the findings of Abdo and Salloum (2017) that soil erosion rates increased with increasing slope and peaked on steep slopes primarily.

4 Conclusions

From 2000 to 2010, the amount of soil erosion was reduced yearly, and the main erosion status in the three periods was very low and low erosion. During the study period, the areas classified as very low erosion were increasing, whilst the areas with low or moderate erosion decreased. Thus erosion intensity decreased over time across the study watershed, with areas of high and very high erosion disappearing by 2010. The soil erosion intensity of terraced fields, sloping land and construction land were mainly very low due to soil and water conservation measures. The erosion intensity of wood land was mainly moderate, and the grassland was mainly low. Soil erosion increases with increasing slope in the range of 0–15°. When the slope is greater than 15°, the change of slope has little effect on the distribution of soil erosion intensity.

The estimates of soil loss using RUSLE are basically consistent with the measured values. Hence the RUSLE model can be used to assess soil erosion conditions in the Zhifanggou watershed and help to target areas in the watershed for maintenance and management procedures to reduce soil erosion risk. A follow-up study should analyse the sensitivity of the model parameters and simplify the parameter requirements of the soil erosion model to effectively simulate soil erosion.

Data availability. 1. The Landsat Thematic Mapper images data of the study area is obtained from the following url: http://glovis.usgs.gov/

2. Daily rainfall data from 1981 to 2004 were obtained from the hydrological stations and precipitation stations from the Institute of Soil and Water Conservation of Pingliang city which is considered confidential.

3. A DEM of 30 m spatial resolution in the study area was obtained from the following url: http://www.gscloud.cn/

Competing interests. The authors declare that they have no conflict of interest.

Special issue statement. This article is part of the special issue "Water quality and sediment transport issues in surface water". It is a result of the IAHS Scientific Assembly 2017, Port Elizabeth, South Africa, 10–14 July 2017.

Acknowledgements. This study is supported by the Fundamental Research Funds for the Central Universities, Ministry of Finance and Education of the People's Republic of China, "Research on the runoff-sediment relationships under climatic change in small watershed in Longdong area of Loess Plateau, China (lzujbky-2012-142)". The authors wish to thank the Institute of Soil and Water Conservation of Pingliang city for providing the original data of runoff-sediment in Zhifanggou watershed.

Edited by: Kate Heal
Reviewed by: Lia Duarte

References

Abdo, H. and Salloum, J.: Mapping the soil loss in Marqya basin: Syria using RUSLE model in GIS and RS techniques, Environ. Earth Sci., 76, 114, https://doi.org/10.1007/s12665-017-6424-0, 2017.

Bai, B., Chen, X. J., and Wang, Y. H.: Interpolation analysis of precipition spatial distribution in Hexiregin of Gansu Province. S10, The development of meteorology and modern agriculture, Chinese meteorological society, Beijing, 2012.

Cai, C. F., Ding, S. W., and Shi, Z. H.: Study of Applying USLE and Geographical Information System IDRISI to Predict Soil Erosion in Small Watershed, J. Soil Water Conserv., 14, 19–24, 2000.

Duarte, L., Teodoro, A. C., Gonçalves, J. A., Soares, D., and Cunha, M.: Assessing soil erosion risk using RUSLE through a GIS open source desktop and web application, Environ. Monit. Assess, 188, 351, https://doi.org/10.1007/s10661-016-5349-5, 2016.

Farhan, Y. and Nawaiseh, S.: Spatial assessment of soil erosion risk using RUSLE and GIS techniques, Environ. Earth Sci., 74, 4649–4669, 2015.

Gao, H. D., Li, Z. B., and Li, P.: The capacity of soil loss control in the Loess Plateau based on soil erosion control degree, Acta Geographica Sinica, 70, 1503–1515, 2015.

Liu, B. Y.: Research report on the development of soil erosion prediction model in the northwest Loess Plateau, Water and water conservation monitoring center in China, R. Beijin, 56–57, 2006.

Liu, B. Y.: Soil erosion prediction model for Chinese studies, Twelfth Conference of the International Soil Conservation, International Soil Conservation Organization, Beijing, 2002.

Liu, B. Y., Nearing, M. A., and Risse, L. M.: Slope gradient effects on soil loss for steep slopes, T. ASAE, 37, 1835–1840, 1994.

Liu, B. Y., Xie, Y., and Zhang, K. L.: Prediction model of soil erosion, Science and Technology Press, M. Beijing, 15–20, 2001.

Lu, H., Gallant, J., Prosser, I. P., Moran, C., and Priestley, G.: Prediction of sheet and rill erosion over the Australian continent, incorporating monthly soil loss distribution, Land and Water Technical Report, CSIRO, Canberra, Australia, 2001.

Meyer, L. D.: Evaluation of the universal soil loss equation, J. Soil Water Conserv., 39, 99–104, 1984.

Pan, Z. K., Huang, J. F., and Wang, F.: Multi range spectral feature fitting for hyperspectral imagery in extracting oilseed rape planting area, Int. J. Appl. Earth Obs., 25, 21–29, 2013.

Renard, K. G. and Ferreira, V. A.: RUSLE model description and database sensitivity, Environ. Qual., 22, 458–466, 1993.

Renard, K. G., Foster, G. R., Weesies, G. A., and Porter, J. P.: RUSLE: Revised universal soil loss equation, Soil Water Conserv., 46, 30–33, 1991.

Renard, K. G., Foster, G. R., Weesies, G. A., McCool, D. K., and Yoder, D. C.: A guide to conservation planning with the revised universal soil loss equation (RUSLE), S. US Department of Agriculture, Agricultural Handbook, No. 703, 1997.

Sharpley, A. N. and Williams, J. R.: EPIC: Erosion Productivity Impact Calculator, Model Documentation, USDA-ARS Tech. Bull. No. 1768, USDA-ARS Grassland, Soil and Water Research Laboratory, Temple, TX, 127, 1990.

Tan, B. X., Li, Z. Y., and Wang, Y. H.: Estimation of Vegetation Coverage and Analysis of Soil Erosion Using Remote Sensing Data for Gui shuihe Drainage Basin, Journal Remote Sensing Technology and Application, 20, 215–220, 2005.

Wischmeier, W. H. and Smith, D. D.: Predicting rainfall erosion losses from cropland east of the Rocky Mountains: guide for selection for practices for soil and water conservation, in: Agriculture handbook. Department of Agriculture, Science and Education Administration, Washington, 1965.

Wischmeier, W. H. and Smith, D. D.: Predicting rainfall erosion losses, a guide to conservation planning, USDA Handbook, 537, US Gov. Print. Off., Washington, DC, 1978.

Yao, X. L., Fu, B. J., Lu, Y. H., and Sun, F. X.: The Soil Moisture Interpolation Method Based on GIS and Statistical Models in Loess Plateau Region, J. Soil Water Conserv., 102, 93–96, 2013.

Zhang, Q.: Study on the effect of terraced fields on the quantitative evaluation of soil erosion, D. Northwestern University, 2016.

Zhao, W. Q., Liu, Y., Luo, M. L., Wang, Y. F., and Lv, Y. H.: Effect of Revegetation on Soil Erosion in Small Watershed of the Loess Plateau, Journal of Soil and Water Conservation, 30, 89–94, 2016.

Proc. IAHS, 377, 19–24, 2018
https://doi.org/10.5194/piahs-377-19-2018
© Author(s) 2018. This work is distributed under
the Creative Commons Attribution 4.0 License.

Regionalising MUSLE factors for application to a data-scarce catchment

David Gwapedza, Andrew Slaughter, Denis Hughes, and Sukhmani Mantel

Institute for Water Research, Rhodes University, P.O. Box 94, Grahamstown, Eastern Cape, South Africa

Correspondence: Andrew Slaughter (a.slaughter@ru.ac.za)
and David Gwapedza (davidgwapedza@gmail.com)

Received: 7 June 2017 – Revised: 8 November 2017 – Accepted: 9 November 2017 – Published: 16 April 2018

Abstract. The estimation of soil loss and sediment transport is important for effective management of catchments. A model for semi-arid catchments in southern Africa has been developed; however, simplification of the model parameters and further testing are required. Soil loss is calculated through the Modified Universal Soil Loss Equation (MUSLE). The aims of the current study were to: (1) regionalise the MUSLE erodibility factors and; (2) perform a sensitivity analysis and validate the soil loss outputs against independently-estimated measures. The regionalisation was developed using Geographic Information Systems (GIS) coverages. The model was applied to a high erosion semi-arid region in the Eastern Cape, South Africa. Sensitivity analysis indicated model outputs to be more sensitive to the vegetation cover factor. The simulated soil loss estimates of $40\,\mathrm{t\,ha^{-1}\,yr^{-1}}$ were within the range of estimates by previous studies. The outcome of the present research is a framework for parameter estimation for the MUSLE through regionalisation. This is part of the ongoing development of a model which can estimate soil loss and sediment delivery at broad spatial and temporal scales.

1 Introduction

Soil erosion is a threat to agriculture and the environment, and water-borne sediment disrupts aquatic ecosystem functionality and compromises the quality of water (Msadala et al., 2010). In addition, reservoir sedimentation is a major offsite impact associated with soil erosion (Kusimi et al., 2015). Soil erosion is therefore a critical environmental problem on a global scale, and is also one of the most important environmental problems facing South Africa, particularly in high soil erosion risk areas such the Eastern Cape Province (Le Roux et al., 2008).

Quantifying the rate of soil loss and sediment delivery as well as identifying major contributing factors is important for the effective and sustainable management of catchments. The development of models that estimate erosion and sediment transport is therefore necessary as models enable planners to gain a better understanding of complex natural processes (Xu, 2002) and the data generated can be used to complement scarce observed sedimentation data.

Erosion modelling has commonly been conducted using the Universal Soil Loss Equation (USLE) (Mishra et al., 2006) or models that are based on a similar conceptual understanding (Rabia, 2012). Modifications of the USLE over the years include the Modified USLE (MUSLE) and the Revised USLE (RUSLE) (Mishra et al., 2006). However, most existing erosion models have been developed for European or North American conditions and may not be reliable or appropriate to represent the dynamics of semi-arid catchments in southern Africa. The spatial and temporal variations of erosion processes in semi-arid catchments complicate the process of simplifying patterns of runoff generation and sediment transfer (Hughes, 2008). This relates to catchments in southern Africa that experience extreme hydrological variability, characterised by low annual precipitation and high evaporative losses. Stored sediment loads in these regions can be abruptly flushed out by sporadic high-intensity storms and flash floods.

Internationally-developed models may also require more observed data for model calibration than are typically available for southern African catchments. To address this problem, Bryson (2015) used MUSLE in conjunction with flow input from the Pitman model (Pitman, 1973) to develop a simple erosion and sediment delivery model (WQSED) to effectively represent the sediment dynamics of South African semi-arid catchments. However, the model is characterised by a large number of parameter requirements. In this context, the aim of the current study was to simplify and reduce the parameter requirements and perform a sensitivity analysis, as well as test the erosion and sediment delivery model. This was aimed at improving/developing a model that can provide simulations of soil loss at appropriate spatial and temporal scales. In this regard, Geographic Information Systems (GIS) analyses of readily-available spatial data were explored for regionalising model parameters.

2 Materials and methods

2.1 Study area

The Tsitsa River catchment is part of the larger Umzimvubu River catchment and is located in the Eastern Cape Province of South Africa. The present study concentrates on the lower quaternary sub catchment of the Tsitsa River catchment, labelled T35E (Fig. 1), which has an area of 492 km^2. The quaternary catchment is the principal water management unit in South Africa and denotes a fourth order catchment in a hierarchal classification system in which a primary catchment is the major unit. Although the Umzimvubu River is noted as the largest undeveloped water resource in South Africa (DWA, 2014), plans are underway to construct a dam in T35E.

The catchment varies considerably in geology, with areas of high elevation along the escarpment consisting of basaltic lava from the Drakensberg Formation (Jurassic), underlain by a stratum of Triassic sandstone and mudstone (Le Roux et al., 2015). Soil depth is limited on steep slopes and gradually deepens towards the foot slopes and floodplain areas due to colluvium and alluvial deposits. The thin soils on steeper slopes become highly erodible when vegetation is degraded (Dollar and Rowntree, 1995), and this progressively worsens as livestock graze on the slopes.

The climate of the area is characterised by a distinct seasonality in rainfall and temperatures. Most rainfall (~ 80 %) occurs during summer (October to March), whereas winter (June to July) is mostly dry. The mean annual rainfall ranges from 625 mm in the low-lying areas to 1415 mm in the mountainous regions (Schulze et al., 2007). Mean monthly temperatures range between 7 °C in winter and 19 °C in summer, with a high variation during the day (Le Roux et al., 2015).

The Tsitsa River catchment is dominated by the grassland biome, and valley bushveld thrives along river channels in the lower reaches of the catchment (Mucina and Rutherford, 2006). The natural vegetation is largely influenced by altitude and burning (Le Roux et al., 2015); therefore, small patches of Afromontane forest occur along drainage lines and ravines where fire has minimal effect.

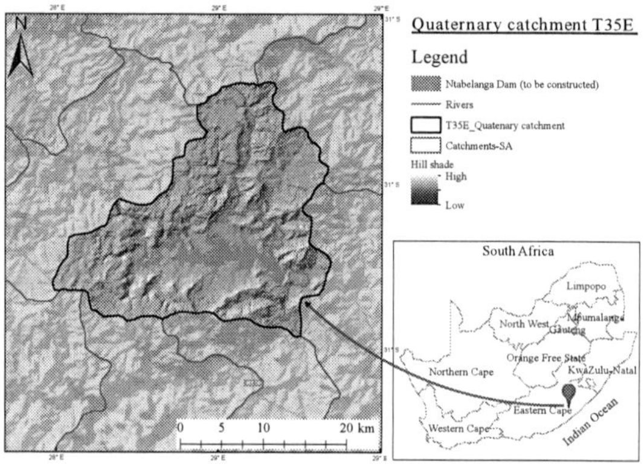

Figure 1. Location of study area: Tsitsa River Catchment, Eastern Cape, South Africa.

2.2 Methodology

The MUSLE was used to estimate soil erosion. The MUSLE is given in the general form of:

$$S_y = a(Qq_p)^b KLSCP, \quad (1)$$

where S_y is sediment yield (t) on a storm basis for the entire catchment, Q is the volume of runoff (m^3), q_p is the peak flow rate ($m^3\,s^{-1}$) and K, L, S, C and P are the soil erodibility ($t\,ha\,h\,MJ^{-1}\,mm^{-1}$), slope length, slope steepness, cover management and soil erosion control practice factors, respectively, similar to the USLE model; parameters a and b are location coefficients. Within the study area for which the equation was developed, the a and b coefficients were 11.8 and 0.56, respectively (Williams and Berndt, 1977). An a priori regionalisation procedure (Kapangaziwiri and Hughes, 2008) was used to estimate K, C and P factors. Regionalisation was adopted to simplify parameter estimation and reduce the user requirements for setting up the model.

In the application of the MUSLE in the present study, the runoff data consists of a monthly discharge record extending from 1920 to 1990. The lack of more recent data records influenced the choice of data that were used within this study. To enable application within MUSLE, the monthly flows from the Pitman Model (Pitman, 1973) were disaggregated to daily. The Pitman Model is one of the most widely used moisture accounting models in southern Africa (Hughes, 2008). Slaughter et al. (2015) present a detailed account of the flow disaggregation method used for the present study. The disaggregated flows were used to obtain the volume (m^3) and peak runoff ($m^3\,s^{-1}$) that drive the MUSLE model (Bryson, 2015).

A sensitivity analysis was performed to determine the changes in model output that occur when different inputs are used in the model (Loucks and Van Beek, 2005). A simple deterministic sensitivity analysis (Benaman, 2003) was used to measure the response of the model output to changes in values of each factor. Minimum and maximum possible values for the study area were used for each factor. According to Loucks and Van Beek (2005), such a range may reflect differences in model outputs between maximum and minimum values for each factor. The model was initially run using catchment parameter values; these were used as the baseline parameters. The next step included routinely running the model using parameters set to the minimum and maximum values respectively. The process involved testing the parameters one at a time so as to evaluate the variations in model output.

The results of the current study were compared to that by Msadala et al. (2010), who predicted sediment yield for South Africa. It is an improvement to the Rooseboom and Lotriet (1992) erosion prediction map for South Africa, and is the largest recent erosion study and a widely-used reference for soil loss in South African catchments. The Le Roux et al. (2015) sediment yield results for the Mzimvubu River catchment were also used for comparison between model outputs as the Tsitsa River is a tributary of the Mzimvubu catchment. The aforementioned studies used the Soil and Water Assessment Tool (SWAT) (Neitsch et al., 2005) and RUSLE (Renard et al., 1997) models to estimate soil loss. The only limitation is that the outputs of the previous studies are at a coarse spatial scale and provided as mean annual soil loss ranges in $\mathrm{t\,km^{-2}}$ and $\mathrm{t\,ha^{-1}}$.

2.3 Determining parameters associated with erodibility

2.3.1 Soil erodibility (K) factor

Soil erodibility refers to the susceptibility of the soil to erosional processes and is dependent on soil characteristics such as structure and texture, which are important determinants of the aggregate soil strength and water infiltration capacity. The K factor is rated on a scale from 0 to 1, with 0 indicating soils with the least susceptibility to erosion, whereas 1 indicates soils which are highly susceptible to soil erosion by water (Schulze et al., 2007).

The soil type distribution for South Africa was obtained from readily-available shapefiles from the South African Atlas of Climatology and Agro-hydrology (Schulze et al., 2007). These data are made available by the Water Research Commission of South Africa (WRC) and contain the distribution of soil types and related K values for soils in South Africa, therefore no empirical soil testing was undertaken. Catchment-specific soil data were extracted from the shapefile by exporting attributes to a Microsoft Excel (2013) spreadsheet. The K factors for various soils were weighted using catchment area to obtain a K factor value that is representative of the entire catchment. This is important because the MUSLE model uses mean values as input for all erodibility parameters.

Table 1. Cover factor (C) for the Ntabelanga Dam catchment.

Land cover/use type	% Area	C	Weighted C
Indigenous Forest	6.8	0.009	0.0006
Woodland/Open bush	3.63	0.012	0.0004
Low shrub land	0.1	0.013	0.0000
Cultivated	12.1	0.37	0.0447
Settlements	7.5	0.1	0.0075
Wetlands	1.59	0.038	0.0006
Grasslands	66.38	0.12	0.0797
Waterbodies	0.1	0.01	0.0000
Bare Ground	0.7	0.45	0.0030
Total	100 %		0.13

2.3.2 Topography (LS) factor

The LS factor was determined using an STRM 30 m digital elevation model (DEM) in a GIS environment. The DEM was clipped to the catchment using a mask extraction tool from the ArcGIS toolbox. The DEM was further conditioned to be depressionless using the "fill sink" command to determine the maximum downhill slope and the flow direction (e.g., Jain and Das, 2010). The slope and flow accumulation were derived from the depressionless DEM. The LS factor map was generated in ArcGIS using the raster calculator (Jain and Das, 2010) by using the LS equation:

$$\mathrm{LS} = \left(\frac{\mathrm{FA}\cdot \mathrm{cellsize}}{22.13}\right)^{0.4} \cdot \left(\frac{\sin(\alpha)0.01745}{0.0896}\right)^{1.3} \qquad (2)$$

where FA is the flow accumulation, and α is the slope gradient in degrees. The cell size is the DEM resolution.

2.3.3 Cover (C) factor

The C factor is a value between 0 and 0.5 that relates to the extent of vegetation cover that protects the soil from erosion in a given catchment (Sadeghi et al., 2013). Cover values closer to 0 indicate dense vegetation cover and reduced erosion output, whereas values close to 0.5 indicate poor vegetation cover. Cover factor values (Table 1) were determined for each land cover type using published guidelines by Wischmeier and Smith (1978), Shinde et al. (2011), Tiruneh and Ayalew (2015), Ranzi et al. (2012) and Jang et al. (2015). Table 1 lists the C factor values for the study area derived used this procedure.

The cover management factor was determined using the National Land Cover data (NLC, 2014). This is the national-scale grid-mapped land cover and land use across South Africa. Catchment-specific cover properties were extracted from the grid by using ArcMap 10.3.1 to clip out catchment-specific data from the NLC map. The attribute table containing land cover categories was exported to a Microsoft Excel (2013) spreadsheet where C factor values published by Wischmeier and Smith (1978) were used to assign C values to

Table 2. Practice factor (P) for the Ntabelanga Dam catchment.

Land cover/use type	% Area	P	Weighted P
Indigenous Forest	0.77	1	0.0077
Thicket/Dense bush	5.99	1	0.0599
Woodland/Open bush	3.63	1	0.0363
Low shrub land	0.05	1	0.0005
Cultivated land	13.35	0.62	0.0824
Settlements	7.50	1	0.0750
Wetlands	1.59	1	0.0159
Grasslands	66.38	1	0.6638
Mines	0.01	1	0.0001
Waterbodies	0.07	1	0.0007
Bare Ground	0.025	1	0.0025
Degraded	0.41	1	0.0041
Total	100 %		0.94

respective land cover classes (Table 1). The C values for particular land cover categories were also verified using C values from recent erosion modelling studies, including Shinde et al. (2011), Tiruneh and Ayalew (2015), Ranzi et al. (2012) and Jang et al. (2015). The mean catchment cover factor was determined by weighting the cover factor against respective percentage catchment area for the different land cover categories (Table 1).

2.3.4 Management practice (P) factor

The management practice factor relates to conservation methods that are implemented to reduce the rate of soil loss from agricultural lands (Tiruneh and Ayalew, 2015). The P factor, which has a value ranging between 0 and 1, refers to management practices that relate to agricultural lands including contour and strip farming. Values closer to 0 indicate good practice and low erosion. The Wischmeier and Smith (1978) table of P values was used to determine P factor values for agricultural land. However, when no conservation was found to be practised, a value of 1 was assigned as the P factor, and all non-agricultural lands were also assigned a P value of 1 if no conservation measure was applied (Tiruneh and Ayalew, 2015; Jang et al., 2015; Luo et al., 2015). Table 2 gives the P factor values for land cover/use types in South Africa.

Using land cover/use maps is a relatively easy and efficient method of determining the P factor (Tiruneh and Ayalew, 2015; Jang et al., 2015; Luo et al., 2015). The P factor for the current study was determined by using land use/land cover maps and a slope map to determine the slope categories in which agriculture is practised within the study catchment. Table 3 displays the erodibility values derived for the study site using GIS and the regionalisation procedure.

Table 3. Modified Universal Soil Loss Equation (MUSLE) erodibility values for the study area.

Catchment	Slope (LS)	Cover (C)	Soil erodibility (K)	Practice (P)
Quaternary T35E	5.3	0.13	0.33	0.94

Table 4. Sensitivity of soil loss simulations (in t × 10^3) to model input parameter values. The P_{min} and P_{max} correspond to the minimum and maximum values of the parameter, respectively.

Parameters	Parameter ranges		Soil loss output	
	Minimum	Maximum	P_{min}	P_{max}
Cover	0.003	0.5	3	530
Soil erodibility	0.03	0.7	12	290
Topography	1	10	25	260
Practice	0.1	1	14	140

3 Results and discussion

3.1 Sensitivity analysis

The result of the sensitivity analysis, summarised in Table 4, shows that the model was more sensitive to the parameter relating to vegetation cover (C) as compared to the other parameters. The minimum and maximum ranges for the C parameter gave the lowest and highest model outputs, respectively. The model output associated with the minimum C (P_{min}) was > 75 % lower compared to model outputs given by setting the other parameters to minimum values. The same trend was noted at the maximum parameter value where the outputs for the maximum C (P_{max}) were > 45 % higher compared to the outputs for other parameters set to their maximums. The model output showed the least sensitivity to the management support practice (P) parameter.

3.2 Model outputs

The simulated results showed that the cumulative amount of soil lost due to erosion in the 492 km^2 T35E quaternary catchment is 137×10^6 t over a 70-year period. The mean annual soil loss is 1.96×10^6 t. This translates to approximately 40 t ha^{-1} yr^{-1} of soil that is lost from the catchment. The result of the current study falls within the range of the findings of Msadala et al. (2010) who estimated that 26–60 t ha^{-1} yr^{-1} is lost from the study area. Le Roux et al. (2015) estimated soil loss for the same area to be 21–50 t ha^{-1} yr^{-1}. The result indicates a high rate of soil loss that is associated with steep slopes (LS, 5.3) and poor vegetation coverage. The C factor value (0.13) indicates that the catchment is mostly covered by grassland, and field studies

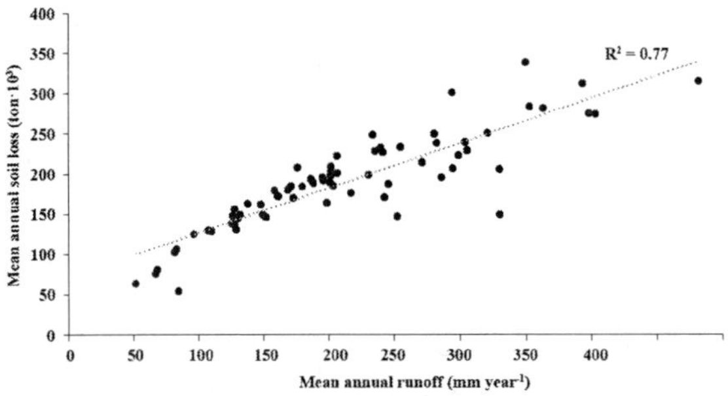

Figure 2. Correlation of runoff and soil loss.

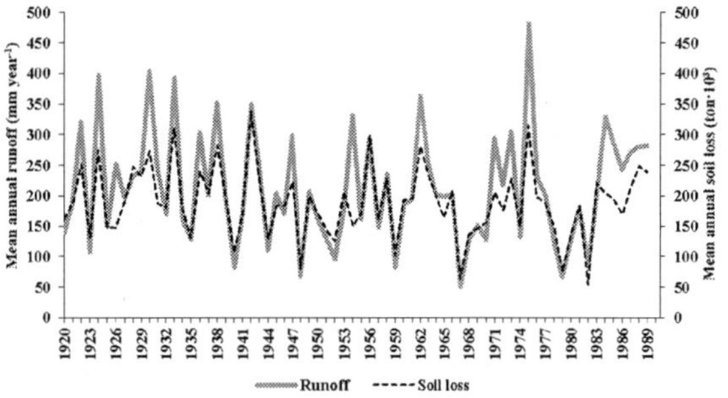

Figure 3. Model output for runoff and soil loss for the Ntabelanga Dam catchment.

conducted in the catchment revealed that the grasslands are mostly degraded by overgrazing and burning.

The rate of soil loss correlated well with runoff (Fig. 2). High flows are typically accompanied by increased soil loss. The modelled time series runoff (Fig. 3) shows the typical "flashiness" associated with arid catchments where periods of dryness are followed by large storm events. This triggers rapid erosion, as displayed by the years 1976 and 1977 (Fig. 3). The model output for soil loss also shows the impact of low flows associated with droughts that affected South Africa. The severe drought period of 1980–1983 (Masih et al., 2014) was associated with low flows and reduced soil loss (Fig. 3).

Based on the experience and findings of the present study, regionalisation of the MUSLE inputs using available GIS datasets reduced data requirements of the model. An effective, simple and low input model is essential for southern African catchments with limited observed data. This supports further development of the soil erosion model. Although the use of readily-available datasets to parameterise the model has been shown to yield reasonable results, a disadvantage of this approach is that temporal variations in vegetation cover have not been considered. Accounting for temporal variations in vegetation cover would likely further improve model performance, and should be considered in the future development of the erosion and sediment transport model.

4 Conclusion and recommendations

The study examined the use of readily-available GIS coverages to derive values for MUSLE factors. An a priori regionalisation procedure was used and values for the LS, C, P and K factors were estimated based on existing GIS data. The outcome of the application of MUSLE under these conditions was reasonable when compared with previous estimates. A sensitivity analysis conducted within this study showed that the model is 50 % more sensitive to the vegetation C factor compared to the other factors. The consideration of temporal changes in the C factor is therefore important. Model outputs correlated well with runoff; an anticipated outcome as the model is typically driven by runoff. A major limitation within the study was that of data, especially observed data, and validating model simulations using previous estimates was not very reliable, although it provides a starting point in a context of data scarcity. For further model development, it is recommended that observed data are collected to enable model outputs to be compared and validated against actual measurements.

Data availability. The data used in this study are available at https://doi.org/10.6084/m9.figshare.5606794 (Gwapedza et al., 2017).

The Supplement related to this article is available online at https://doi.org/10.5194/piahs-377-19-2018-supplement.

Competing interests. The authors declare that they have no conflict of interest.

Special issue statement. This article is part of the special issue "Water quality and sediment transport issues in surface water". It is a result of the IAHS Scientific Assembly 2017, Port Elizabeth, South Africa, 10–14 July 2017.

Acknowledgements. This research was funded by the Water Research Commission (WRC) South Africa. The data were provided by the Department of Water and Sanitation, Department of Environment Affairs, USGS explorer and the WRC. They are thanked for making the GIS data and flow data readily available and free to access.

Edited by: Kate Heal
Reviewed by: Seyed Hamidreza Sadeghi and Ju Qian

References

Benaman, J.: A Systematic Approach to Uncertainty Analysis for a Distributed Watershed Model, PhD Thesis, School of Civil and Environmental Engineering, Cornell University, Ithaca, NY, 2003.

Bryson, L.: An erosion and sediment delivery model for semi-arid catchments, MS Thesis, Rhodes University, Grahamstown, 2015.

Dollar, E. S. J. and Rowntree, K. M.: Sediment sources, hydro climatic trends and geomorphic response in a mountainous catchment, north Eastern Cape, South Africa, S. Afr. Geogr. J., 77, 21–32, 1995.

DWA: Umzimvubu Water Project, http://www.dwa.gov.za/mzimvubu/documents/announcement/BackgroundInformationDocument(English).pdf (last access: 8 December 2017), 2014.

Hughes, D. A.: Simulating the hydrology and total dissolved solids (TDS) of an ephemeral river in South Africa for environmental water requirement determinations, River Res. Appl., 24, 1–11, 2008.

Jain, M. K. and Das, D.: Estimation of Sediment Yield and Areas of Soil Erosion and Deposition for Watershed Prioritization using GIS and Remote Sensing, Water Resour. Manage., 24, 2091–2112, 2010.

Jang, C., Shin, Y., Kum, D., Kim, R., Yang, J. E., Kim, S. C., and Jung, Y.: Assessment of soil loss in South Korea based on land-cover type, Stoch. Environ. Res. Risk A., 29, 2127–2141, 2015.

Kapangaziwiri, E., & Hughes, D.: Towards revised physically based parameter estimation methods for the Pitman monthly rainfall-runoff model. Water SA, 34(2), 183–192, 2008.

Kusimi, J. M. andAttua, E. M.: Soil Erosion and Sediment Yield Modelling in the Pra River Basin of Ghana using the Revised Universal Soil Loss Equation (RUSLE), Ghana J. Geogr., 7, 38–57, 2015.

Le Roux, J. J., Morgenthal, T. L., Malherbe, J., Pretorius, D. J., and Sumber, P. D.: Water erosion prediction at a national scale for South Africa, Water SA, 34, 305–314, 2008.

Le Roux, J. J., Barker, C. H., Weepener, H. L., Van den Berg, E. C., and Pretorius, S. N.: Sediment yield modelling in the Umzimvubu river catchment, WRC Report No. 2243/1/15, Water Research Commission, Pretoria, South Africa, 2015.

Loucks, D. P. and Van Beek, E.: Water Resources Systems Planning and Management. An Introduction to Methods, Models and Applications, UNESCO Publishing, Paris, France, 2005.

Luo, Y., Yang, S., Liu, X., Liu, C., Zhang, Y., Zhou, Q., and Dong, G.: Suitability of revision to MUSLE for estimating sediment yield in the Loess Plateau of China, Stoch. Environ. Res. Risk A., 30, 379–394, 2015.

Masih, I., Maskey, S., Mussá, F. E. F., and Trambauer, P.: A review of droughts on the African continent: a geospatial and long-term perspective, Hydrol. Earth Syst. Sci., 18, 3635–3649, https://doi.org/10.5194/hess-18-3635-2014, 2014.

Mishra, S. K., Tyagi, J. V., Singh, V. P., and Singh, R.: SCS-CN-based modelling of sediment yield, J. Hydrol., 324, 301–322, 2006.

Msadala, V., Gibson, L., Le Roux, J. J., Rooseboom, A., and Basson, G. R.: Sediment Yield Prediction for South Africa: 2010 Edition, WRC report 1765/1/10, Water Research Commission, Pretoria, South Africa, 2010.

Mucina, L. and Rutherford, M. C.: The vegetation of South Africa, Lesotho and Swaziland, Strelitzia 19, South African National Biodiversity Institute, Pretoria, South Africa, 2006.

Neitsch, S. L., Arnold, J. G., Kiniry, J. R., and Williams, J. R.: Soil and Water Assessment Tool – Theoretical Documentation, Version 2005, Texas Water Resources Institute, Texas, 2005.

NLC – National Land Cover: 2014 South African National Land-Cover, retrieved from Biodiversity GIS: http://bgis.sanbi.org/DEA_Landcover (last access: 8 December 2017), 2014.

Pitman, W. V.: A Mathematical Model for Generating River Flows from Meteorological Data in South Africa, Report No. 2/73, Hydrological Research Unit, University of the Witwatersrand, Johannesburg, South Africa, 1973.

Rabia, A. H.: Mapping Soil Erosion Risk Using RUSLE, GIS and Remote Sensing Techniques, in: The 4th International Congress of ECSSS, EUROSOIL, 2–6 July 2012, Bari, Italy, 2012.

Ranzi, R., Le, T. H., and Rulli, M. C.: A RUSLE approach to model suspended sediment load in the Lo River (Vietnam): Effects of reservoirs and land use changes, J. Hydrol., 422–423, 17–29, 2012.

Renard, K., Foster, G., Weesies, G., McCool, D., and Yoder, D.: Predicting soil erosion by water: a guide to conservation planning with the Revised Universal Soil Loss Equation (RUSLE), Agricultural Handbook No. 703, United States Department of Agriculture, Washington, D.C., 1997.

Rooseboom, A. and Lotriet, N. H.: The new sediment yield map for southern Africa. Erosion and Sediment Transport Monitoring Programmes in River Basins, in: Proceedings of the IAHS Symposium, 24–28 August 1992, Oslo, Norway, 210 pp., 1992.

Sadeghi, S. H. R., Gholami, L., Khaledi Darvishan, A., and Saeidi, P.: A review of the application of the MUSLE model worldwide, Hydrolog. Sci. J., 59, 365–375, 2013.

Schulze, R. E., Maharaj, M., Warburton, M. L., Gers, C. J., Horan, M. J. C., Kunz, R. P., and Clark, D. J.: South African atlas of climatology and agrohydrology, RSA, WRC Report 1489, Water Research Commission, Pretoria, 2007.

Shinde, V., Sharma, A., Tiwari, K. N., and Singh, M.: Quantitative Determination of Soil Erosion and Prioritization of Micro-Watersheds Using Remote Sensing and GIS, J. Indian Soc. Remote Sens., 39, 181–192, 2011.

Slaughter, A. R., Retief, D. C. H., and Hughes, D. A.: A method to disaggregate monthly flows to daily using daily rainfall observations: model design and testing, J. Hydrol., 4, 153–171, 2015.

Tiruneh, G. and Ayalew, M.: Soil loss estimation using geographical information system in Enfraz watershed for soil conservation planning in Highlands of Ethiopia, Int. J. Agril. Res. Innov. Tech., 5, 21–30, 2015.

Williams, J. R. and Berndt, H. D.: Sediment yield based on watershed hydrology, T. Soc. Agr. Eng., 20, 1100–1104, 1977.

Wischmeier, W. H. and Smith, D. D.: Predicting rainfall erosion losses: a guide to conservation planning, Agricutural Handbook no. 537, United States Department of Agriculture, Washington, D.C., 1978.

Xu, C.: Textbook of Hydrologic Models, in: Vol. 72, Uppsala University Department of Earth Sciences Hydrology, Uppsala, 2002.

Proc. IAHS, 377, 25–33, 2018
https://doi.org/10.5194/piahs-377-25-2018
© Author(s) 2018. This work is distributed under
the Creative Commons Attribution 3.0 License.

Water quality modelling of an impacted semi-arid catchment using flow data from the WEAP model

Andrew R. Slaughter and Sukhmani K. Mantel

Institute for Water Research, Rhodes University, Grahamstown, South Africa

Correspondence: Andrew R. Slaughter (a.slaughter@ru.ac.za)

Received: 6 June 2017 – Accepted: 14 August 2017 – Published: 16 April 2018

Abstract. The continuous decline in water quality in many regions is forcing a shift from quantity-based water resources management to a greater emphasis on water quality management. Water quality models can act as invaluable tools as they facilitate a conceptual understanding of processes affecting water quality and can be used to investigate the water quality consequences of management scenarios. In South Africa, the Water Quality Systems Assessment Model (WQSAM) was developed as a management-focussed water quality model that is relatively simple to be able to utilise the small amount of available observed data. Importantly, WQSAM explicitly links to systems (yield) models routinely used in water resources management in South Africa by using their flow output to drive water quality simulations. Although WQSAM has been shown to be able to represent the variability of water quality in South African rivers, its focus on management from a South African perspective limits its use to within southern African regions for which specific systems model setups exist. Facilitating the use of WQSAM within catchments outside of southern Africa and within catchments for which these systems model setups to not exist would require WQSAM to be able to link to a simple-to-use and internationally-applied systems model. One such systems model is the Water Evaluation and Planning (WEAP) model, which incorporates a rainfall-runoff component (natural hydrology), and reservoir storage, return flows and abstractions (systems modelling), but within which water quality modelling facilities are rudimentary. The aims of the current study were therefore to: (1) adapt the WQSAM model to be able to use as input the flow outputs of the WEAP model and; (2) provide an initial assessment of how successful this linkage was by application of the WEAP and WQSAM models to the Buffalo River for historical conditions; a small, semi-arid and impacted catchment in the Eastern Cape of South Africa. The simulations of the two models were compared to the available observed data, with the initial focus within WQSAM on a simulation of instream total dissolved solids (TDS) and nutrient concentrations. The WEAP model was able to adequately simulate flow in the Buffalo River catchment, with consideration of human inputs and outputs. WQSAM was adapted to successfully take as input the flow output of the WEAP model, and the simulations of nutrients by WQSAM provided a good representation of the variability of observed nutrient concentrations in the catchment. This study showed that the WQSAM model is able to accept flow inputs from the WEAP model, and that this approach is able to provide satisfactory estimates of both flow and water quality for a small, semi-arid and impacted catchment. It is hoped that this research will encourage the application of WQSAM to an increased number of catchments within southern Africa and beyond.

1 Introduction

At a global level, there is an increasing trend of water quality deterioration (Verhoeven et al., 2006). South Africa, which is typical of developing countries with a semi-arid environment, suffers from water quality pollution resulting from poor infrastructure maintenance and management (Adler et al., 2007), the main manifestations of which are inputs of unacceptably high loads of nutrients into rivers from Waste Water Treatment Plants (WWTPs), leading to eutrophic conditions. Besides eutrophication, major water quality problems experienced by South African rivers include poor microbial water quality (Britz and Sigge, 2012) and acid mine drainage (Simate and Ndlovu, 2014).

South Africa has a relatively strong history of water quantity management, with region-specific hydrology and systems models being developed that are routinely used in water resources management. These include the Pitman (1973) and the Agricultural Catchments Research Unit (ACRU) models (Schulze, 1989) (hydrological models) and the Water Resources Yield (WRYM) (Basson et al., 1994) and Water Resources Modelling Platform (WReMP) (Mallory et al., 2011) systems models, modelling natural and human-altered flow respectively. In contrast, water quality modelling tools specific for application to southern African catchments are poorly developed, and in South Africa, this has led to the water quality component of water resources management being loosely connected to the water quantity component. Although some internationally-developed water quality models have been applied to South African surface waters within scientific studies (e.g. Gorgens and de Clercq, 2006; Piesold et al., 2007), the adoption of a routinely-used water quality model within management agencies in South Africa has been hampered by the lack of observed data with which to calibrate these models and insufficient technical expertise to accurately set up these models. This led to the development of the Water Quality Systems Assessment Model (Slaughter et al., 2012, 2015, 2017; Hughes and Slaughter, 2016; Slaughter and Mantel, 2016; Slaughter, 2017; Slaughter and Mantel, 2017a), which specifically links to the routinely-used systems models in South Africa, namely the WRYM and WReMP models, and subscribes to an approach of requisite simplicity (Stirzaker et al., 2010) by only representing processes that explain the majority of variation of observed water quality data.

The approaches used and the validity of the WQSAM model have been validated in previous studies, including the monthly-to-daily flow disaggregation (Slaughter et al., 2015; Hughes and Slaughter, 2016) and the water quality modelling processes (Slaughter et al., 2012, 2017; Slaughter and Mantel, 2016; Slaughter, 2017), which include simulations of salinity as total dissolved solids (TDS), nutrients (nitrates plus nitrites, phosphates and ammonia) and microbial water quality, illustrating that the approach adopted by WQSAM is appropriate for modelling water quality in South Africa under conditions of limited observed data and management capacity.

The limitations to water quality modelling in South Africa, as a semi-arid developing country, are likely common to most countries within southern Africa, and possibly other regions globally. A broader aim should therefore be the facilitation of the application of WQSAM within other semi-arid developing regions. This would further refine and validate the water quality processes used in WQSAM, and also provide a useful water quality management tool in those regions. The plausibility of this development is constrained by WQSAM's dependence on the systems models, the WReMP and the WRYM, as although the approach used by WQSAM of connecting with these models is useful for management from a South African perspective, this approach also limits the use of WQSAM to South Africa, or in a few catchments in southern Africa where these systems models may have been applied. The Water Evaluation and Planning (WEAP) model (Sieber and Purkey, 2007) is an easy-to-use water accounting model that was developed by the Stockholm Environment Institute, and has been applied to catchments in many developing countries (e.g. Levite et al., 2003; Hollermann et al., 2010; Hamlat et al., 2013; Dimova et al., 2014; Li et al., 2015). Although the WEAP model has been proven to be enormously useful for modelling water quantity in data-poor regions, the water quality simulation capabilities of the model for long-term catchment-level management remain limited. A valid approach to increasing the utility of WQSAM outside of South Africa would be to adapt WQSAM to take as input, flow output of the WEAP model.

The aim of the current study is to describe the process of adapting the WQSAM model to accept flow data output of the WEAP model, and by application of the model to a case study catchment, to illustrate the validity of the approach.

2 Materials and methods

2.1 Study area

The Buffalo River in the Eastern Cape, South Africa was used as a case study catchment for the current study (Fig. 1). This river is relatively short (125 km from headwaters to sea), with a maximum order (Strahler, 1957) of 4 and few tributaries. Although rainfall occurs year round, most rainfall occurs during summer, and there is a steep rainfall gradient between the upper (1500–2000 mm) and the middle (500–625 mm) reaches. Four impoundments are situated on the river: Maden and Rooikrans dams in the upper catchment and the larger Laing Dam and Bridle Drift Dam in the middle and lower catchment, respectively (Fig. 1). It must be emphasised that the current study modelled water quality up to Laing Dam (middle catchment). The natural salinity of the river is relatively high as the geology of the region is dominated by marine sediments of the Beaufort Series. Human activities affecting water quality include some commercial farming in

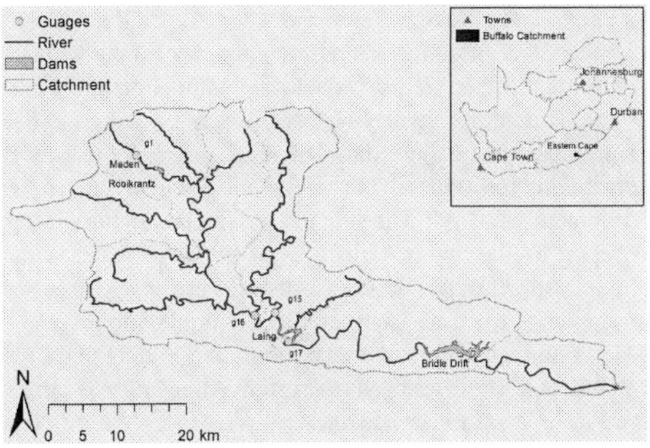

Figure 1. Map of Buffalo River Catchment within the Eastern Cape, South Africa. Water quality monitoring points used in the study for calibration of the models are indicated: g1 – Maden Dam; g16 – R20D; g15 – Yellowwoods River; g17 – Laing Dam.

the upper catchment, and once water moves through the middle catchment, there is a pronounced deterioration of water quality due to pollution from the urban/industrial complex of King Williams Town and Zwelitcha (see Fig. 1). The pollution of the middle catchment mainly manifests as eutrophication, resulting from nutrient inputs from various WWTPs. This eutrophication is mainly manifested in Laing Dam, resulting in blooms of Water Hyacinth (*Eichhornia crassipes*) and Microcystis.

2.2 Models used

The Water Evaluation and Planning (WEAP) Model (Sieber and Purkey, 2007) was used to simulate monthly flows within the catchment. WEAP is an "off-the-shelf" water accounting model, used for simulating natural hydrology through a rainfall-runoff function, and is also capable of simulating human-altered flow through simulating human use extractions and return flows. WEAP has been widely applied due to its ease of use. Although WEAP is a good systems model for simulating water quantity, the model incorporates only very rudimentary water quality simulation functionality, with conservative water quality variables modelled through dilution and non-conservative variables modelled though a single globally-applicable degradation coefficient. Although WEAP does allow integration of the QUAL2K model (Pelletier et al., 2006) for modelling water quality, this approach is not ideal as QUAL2K is a highly detailed water quality model with a high spatial and temporal resolution, meant for detailed modelling of small stretches of river, whereas WEAP would typically be applied at a catchment level for long-term simulations of water quantity.

The Water Quality Systems Assessment Model (WQSAM) was used to simulate water quality. As mentioned in the Introduction, WQSAM was designed to link directly to systems models routinely used in South African water resource management; the flows generated by the systems models drive the water quality simulations in WQSAM. At the conceptual level, WQSAM can be represented by several levels (tiers) (Slaughter et al., 2012). The functionality of the first tier would be to facilitate the communication of WQSAM with the systems model in allowing the input of flow data. WQSAM is run within the Spatial And Time Series Information Modelling (SPATSIM) modelling framework (Hughes, 2004). Typically, the systems models (WRYM or WReMP) generate monthly flows, as monthly flows are regarded as adequate for the needs of long-term water quantity management in South Africa. Since water quality is strongly affected by transient events (Britton et al., 1993), such as rainfall-runoff events, WQSAM is operated on a daily time step. The monthly flows are therefore disaggregated to daily within WQSAM, represented by the second tier of the model. A detailed description of the disaggregation method is available in Slaughter et al. (2015), and is further explored in Hughes and Slaughter (2015). The third tier of WQSAM represents the separation of incremental flow into the three flow fractions: surface water flow, interflow and ground water flow. This process is achieved through the simple statistical baseflow separation method of Hughes et al. (2003). The flow separation is of particular importance for simulation of non-point source loads, as water quality signatures (concentrations) are assigned to the flow fractions, the values of which are guided by the predominant land covers in the sub-catchments. The disaggregation of monthly cumulative flow into daily, including return flows, abstractions and reservoir releases, is represented by the fourth tier and has been described in Hughes and Slaughter (2016). All water quality simulation, including that of nutrients, salinity, water temperatures and microbial water quality, is represented by the final tier of the model, and includes processes such as decomposition, chemical speciation and algal uptake (Slaughter et al., 2012).

Within the current study, the model functionality of the WQSAM model was updated to read in monthly flow output from the WEAP model. The updated part of the WQSAM model is conceptually represented by the first tier. Briefly, WEAP outputs simulations in various forms, including tables and graphs, but usefully for the present study, the unprocessed monthly simulations can be output to an excel spreadsheet as a comma-delimited file. The WQSAM functionality represented as tier one to read in monthly flow data was therefore updated to provide the option of reading in comma-delimited files. The user interface of WQSAM to allow this option is shown in Fig. 2.

Information on dams on the system, such as dam capacity, are written to a text file, and would be input into WQSAM in the field represented as "1" in Fig. 2. The "YldStats.out" file is a standard output file of the Water Resources Modelling Platform (WReMP) (Mallory et al., 2011), and includes information on outflows from subcatchments and reservoirs

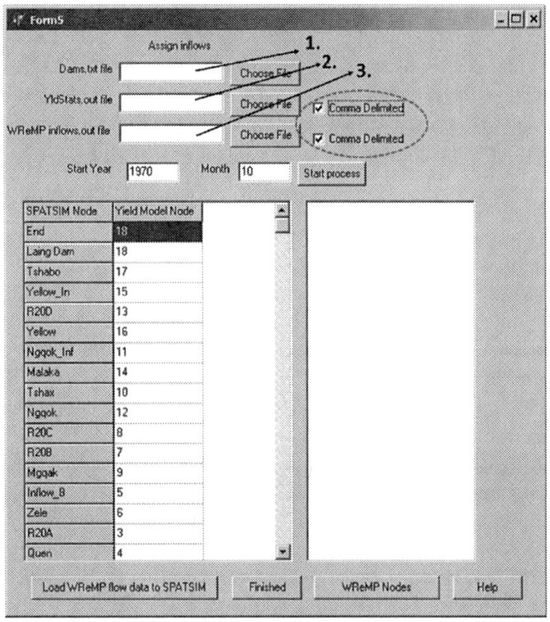

Figure 2. User interface of the Water Quality Systems Assessment Model (WQSAM) (Slaughter et al., 2012) for reading in flow outputs of the systems models.

and reservoir water levels at monthly time steps, and is input into WQSAM in the field represented as "2" in Fig. 2. The "inflows.out" file is a standard output file of the Water Resources Modelling Platform (WReMP) (Mallory et al., 2011), and includes information on both incremental and cumulative inflows to subcatchments and reservoirs at monthly time steps, and is input into WQSAM in the field represented as "3" in Fig. 2. By checking the check boxes indicated in the circle in Figure 2, the model functionality to read in the two WReMP files as comma delimited files is activated, thereby allowing the model to read in flow outputs from the Water Evaluation and Planning (WEAP) model (Sieber and Purkey, 2007). It must be emphasised that WQSAM cannot at this stage simply ready in the flow output files of WEAP in their raw form. The flow outputs of WEAP are reformatted and separated into the flows corresponding to the YldStats.out and inflows.out files mentioned above. This can be a time-consuming process.

Once monthly flows are read into the WQSAM model, all other processes are run without any changes: monthly flows are disaggregated to daily, flows are fractioned and water quality is simulated.

2.3 Model setup

WEAP was used to simulate future water quantity within the Buffalo River for the period 1979–2000 on a monthly time step. More information on the WEAP model setup and water use is available in Mantel et al. (2015). Briefly, historical rainfall and evaporation/temperature data were obtained from the South African WR2005 database (Middleton and Bailey, 2008). The natural hydrology was generated within WEAP using these data using a built-in rainfall-runoff module. By considering reservoirs in the catchment as well as current water demands, WEAP was able to simulate altered hydrology. A report by the South African Department of Water Affairs and Forestry (DWAF, 2008) was used to estimate current water demands for the system, and for simplification, were divided into three demand areas: the upper, middle and lower catchments. The present day (current) demand was considered to be a stationary demand in WEAP for the modelled years 1971–2000. The water losses due to evapotranspiration by invasive plants was also considered in the WEAP application for the Amatole system. Reticulation losses were also considered. Calibration of stream flows was achieved using observed data for 11 gauging stations on the system.

The WEAP monthly modelled flows were output to comma-delimited files and reformatted into the YldStats.out and Inflows.out files, following which the data were input into WQSAM through the user interface represented in Fig. 2. Monthly incremental flows were disaggregated to daily using observed daily rainfall from the rainfall database by Lynch (2004). Water temperature was simulated using the daily observed air temperature database by Schulze and Maharaj (2004). Water quality data collected by the South African Department of Water and Sanitation (DWS) were used to calibrate the model.

3 Results

3.1 Inputs of flow data from the WEAP model into WQSAM

The study by Mantel et al. (2015) found that the WEAP model was in general able to represent the observed flow of the Buffalo River catchment for historical conditions. To illustrate whether the WQSAM model was able to input the monthly flow data generated by the WEAP model, we chose two example sub-catchments, R20B and Yellowwoods, for which the WEAP model simulated flow relatively inaccurately and relatively accurately, respectively. For the R20B subcatchment (Fig. 3a), the time series of simulated monthly flows generated by the WEAP model (top graph) appears to be too high compared to observed monthly flows. Once the flows were input into the WQSAM model, the process of monthly-daily disaggregation produced daily flows (middle graph) that were similarly too high compared to observed daily flows. The inaccuracies in the monthly and daily flows are illustrated in the flow duration curves (lower graph), where the comparisons between simulated and observed flows for monthly and daily time steps generated Nash-Sutcliffe Efficiency (NSEs) (Nash and Sutcliffe, 1970) of 0.16 and 0.1, respectively. For the Yellowwoods River sub-catchment, the WEAP model generated monthly flows that were more representative of the observed flows, as can be seen in the top graph of Fig. 3b. The monthly-daily disaggregation in WQSAM produced daily flows that were similarly representative of the observed daily flows (middle graph). The flow duration curves within the lower graph show that the comparisons between simulated and observed flows for monthly and daily time steps generated NSEs of 0.93 and 0.98, respectively.

3.2 Water quality modelling in WQSAM

Observed water quality data were available for Maden Dam, R20D, the Yellowwoods River and Laing Dam. Maden Dam is at the headwaters of the catchment, and water quality is therefore relatively good. TDS, NO_3-N + NO_2-N, NH_4-N and PO_4-P for Maden Dam ranges between 30 to 130 mg L^{-1}, 0 to 1.1 mg L^{-1}, 0 to 1.4 mg L^{-1} and 0 to 1.0 mg L^{-1}, respectively. To achieve calibration to the observed data for Maden Dam, signatures of surface water, interflow and groundwater flow for TDS were set at 30, 30 and 80 mg L^{-1}, respectively. For the nutrients, only the surface water signatures were set, with values of 0.2, 0.2 and 0.1 mg L^{-1} for NO_3-N + NO_2-N, NH_4-N and PO_4-P, respectively. Minimal algal growth was simulated within the dam. The water quality of the R20D catchment shows a significant deterioration compared to the headwaters as this point is within the middle catchment and is affected by return flows and runoff from urban areas. TDS in the catchment shows a sharp rise, with values ranging between 100 and 5000 mg L^{-1}. Nutrients also show a sharp rise, with highest concentrations of NO_3-N + NO_2-N, NH_4-N and PO_4-P of approximately 46, 14 and 6 mg L^{-1}, respectively. To achieve calibration for TDS, surface water, interflow and groundwater flow concentrations were set to 100, 500 and 1200 mg L^{-1}, respectively. For the nutrients, surface flow concentrations were set to high levels to achieve calibration, with values 1, 5 and 0.5 mg L^{-1} for NO_3-N + NO_2-N, NH_4-N and PO_4-P, respectively. The concentrations of return flow for the nutrients were also set at high levels to represent sewage return flow into the river at that point, with values of 10, 0.5 and 1.5 mg L^{-1} for NO_3-N + NO_2-N, NH_4-N and PO_4-P, respectively. The water quality of the Yellowwoods River is also fairly compromised compared to the upper catchment, as the river receives sewage return flow input and runoff from informal settlements. TDS within the river ranges between 100 and 1300 mg L^{-1}. Nutrients are also fairly high, with concentrations of NO_3-N + NO_2-N, NH_4-N and PO_4-P being as high as 5.5, 10 and 3 mg L^{-1}, respectively. To achieve calibration to the observed TDS, signatures of 100, 500 and 1600 mg L^{-1} were assigned to surface flow, interflow and groundwater flow, respectively. The surface flow concentrations for nutrients were set high to achieve calibration, with values of 5, 1.5 and 1 mg L^{-1} for of NO_3-N + NO_2-N, NH_4-N and PO_4-P, respectively.

Table 1. Nash-Sutcliffe Efficiency (NSE) (Nash and Sutcliffe, 1970) values calculated for the Water Quality Systems Assessment Model (WQSAM) (Slaughter et al., 2012) simulations compared against historical observed data for various water quality variables for various subcatchments/reservoirs of the Buffalo River catchment.

Subcatchment/ Reservoir	TDS	NO_3-N +NO_2-N	NH_4-N	PO_4-P
Maden Dam	0.93	0.84	0.37	0.93
R20D	0.64	0.93	0.93	0.62
Yellowwoods	0.83	0.88	0.14	0.47
Laing Dam	0.28	0.56	10.15	−1.23

Return flow concentrations were set fairly low, with values of 0.5, 0.1 and 0.1 for NO_3-N + NO_2-N, NH_4-N and PO_4-P, respectively. This indicates that non-point sources have a bigger impact on the Yellowwoods River than point sources. Laing Dam appears to act as a water quality sink, as although the water quality is generally compromised, it is nevertheless an improvement compared to the water quality of the dam inflow. Observed TDS ranges between 150 to 500 mg L^{-1}. The highest observed concentrations of nutrients are approximately 3, 1.5 and 1.5 mg L^{-1} for NO_3-N + NO_2-N, NH_4-N and PO_4-P, respectively.

The NSEs generated by comparing model simulations of water quality in WQSAM to observed data are shown in Table 1. The simulations of TDS appear to be most representative of observed data. The model simulations of water quality for Laing Dam represented the most inaccurate representations of the observed data. The NSEs for the other subcatchments/reservoirs ranged between 0.64–0.93, 0.84–0.93, 0.14–0.93 and 0.47–0.93 for TDS, nitrate plus nitrite, ammonia and phosphate, respectively.

4 Discussion

4.1 Model results

Figure 3 shows that WEAP generated a relatively poor representation of observed monthly flow in the R20B subcatchment (see Fig. 3a), with the NSE value being 0.16. This monthly flow was input into WQSAM and disaggregated from monthly to daily, and then compared to the daily observed flow, generating an NSE of 0.10. Therefore, the poor monthly flows generated by the WEAP model were carried over into the WQSAM model, shown by the observed monthly versus simulated monthly and the observed daily versus simulated daily flow duration curves, respectively. In contrast, the monthly flows generated by the WEAP model for the Yellowwoods River were a good representation of the observed data, generating an NSE value of 0.93.

This good representation of flow was carried through to the WQSAM model, and subsequent the monthly-daily flow dis-

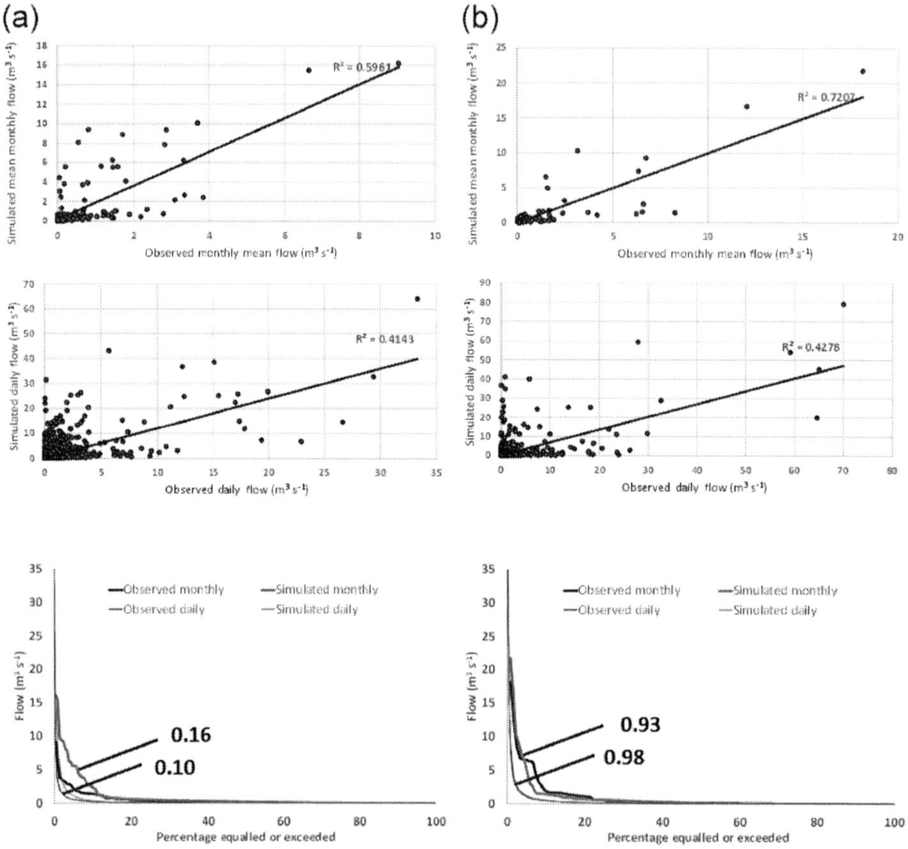

Figure 3. Observed and simulated flow for: **(a)** R20B and; **(b)** Yellowwoods on the Buffalo River in the Eastern Cape for the period 1971–2000. The top graph shows the *X–Y* scatter plot of the observed monthly time series versus WEAP (Sieber and Purkey, 2007) model simulations of monthly flow, the middle graph shows the *X–Y* scatter plot of the daily observed flow versus daily simulated flow generated through monthly-daily disaggregation within the WQSAM (Slaughter et al., 2012) model and the lower graph shows daily observed versus daily simulated and monthly observed versus monthly simulated flow represented as duration curves, with the Nash-Sutcliffe efficiencies (Nash and Sutcliffe, 1970) shown between each comparison.

aggregation, a comparison between daily observed and daily simulated flow yielded an NSE of 0.98. The results shown in Fig. 3 suggest that flow was successfully input from WEAP into WQSAM; therefore, the additional functionality built into WQSAM to input WEAP-generated flows appears to be effective.

Water quality simulations within WQSAM obtained mixed results. It is evident that the model struggled to simulate water quality within Laing Dam. The following summary of results therefore excludes the results for Laing Dam. Unsurprisingly, model simulations of the conservative variable TDS were relatively representative of observed data, with NSE values ranging between 0.64–0.93. Simulations of nitrates plus nitrites were also relatively representative of observed data, with NSE values ranging between 0.56–0.93. Simulations of ammonium and phosphate were mixed, and although some good results were obtained, some poor results were also obtained. The poor model NSE results for ammonium within Maden Dam as well as the Yellowwoods River and phosphates within the Yellowwoods River could be related to the scarcity of observed data as well as a few observations of very high concentrations, which could possibly be due to measurement or data capture errors: in general, the model simulations were representative of the observed data except for a few extreme spikes in the observed data. WQSAM models TDS at this stage primarily through dilution. The relatively good simulation results obtained for TDS by the model indicate that in general, the simulated flows used in the model were representative of the observed flows, which further confirms that the WQSAM model was successfully able to input the monthly flows generated by the WEAP model. The mixed results obtained for nutrient simulations by the model indicate model uncertainties and possible inadequacies in representing observed water quality that are probably not due to incorrect flow, but rather possibly due to incorrect or insufficient representation of certain water quality processes. It is also possible that the extreme spikes in the observed water quality data, particularly for nutrients, may be related to measurement or data capture errors. Since there are few observed data, the fact that WQSAM was in most cases not able to represent these spikes, sometimes led to poor NSE estimates.

4.2 The benefit of modelling water quality in WQSAM

As already mentioned in the Introduction, WQSAM aims to represent the most important water quality processes that explain the majority of variation in observed flows. In this way, relative simplicity of the model can be maintained, thereby also limiting the observed data required to calibrate the model. Many systems models, such as the WEAP application to the Buffalo River Catchment (Mantel et al., 2015) and all model setups of the WReMP and WRYM models in South Africa, are at a monthly time step. This is because simulations of water quantity at a monthly time step are generally sufficiently accurate for long-term water quantity management in South Africa. Simulation at a monthly time step would not be sufficiently accurate for water quality, as water quality is typically driven by transient events, such as rainfall-runoff events. An important advantage of WQSAM is its ability to integrate with the monthly-time-step systems models, generating water quality at a daily time step, thereby utilising the existing systems model setups and avoiding complex daily-time-step systems models. An example of the benefit of modelling at a daily time step as opposed to a monthly time step can be seen in Fig. 4a, which shows the frequency distributions of simulations of TDS at a daily time step and monthly time step by WQSAM and the WEAP model, respectively, for the R20D subcatchment in the Buffalo River Catchment, along with the frequency distribution of observed TDS. The simulation of water quality within the Buffalo River by WEAP was conducted as part of the study by Slaughter et al. (2016). It is evident that the simulation of TDS by WQSAM shows more variation, and is more representative of the observed daily data, whereas the simulation by WEAP is of monthly averages, and does not represent the full variation of the observed data. Obviously, it would be of greater use for water management to be aware of the full variation of a particular water quality variable, so as to estimate the real risk of exceeding certain management thresholds, associated with particular management actions. As can be seen in Fig. 4b, which shows the X–Y plot of WQSAM-simulated daily TDS versus observed TDS, a relatively poor correlation is obtained. It must be considered that WQSAM is not designed to accurately match observed water quality observations at a daily time scale, but rather to generate a representative frequency distribution of long-term water quality observations.

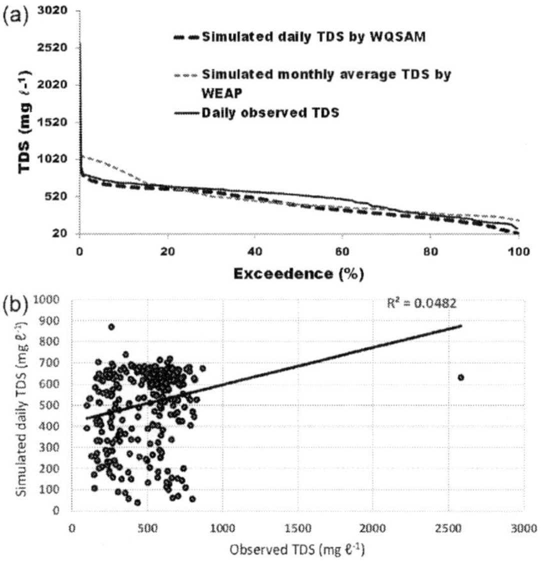

Figure 4. Frequency distribution of observed total dissolved solids (TDS), modelled daily TDS by the Water Quality Systems Assessment Model (WQSAM) and modelled average monthly TDS by the monthly-time-step Water Evaluation and Planning (WEAP) model (Sieber and Purkey, 2007) for the R20D catchment on the Buffalo River, Eastern Cape, South Africa. **(b)** X–Y plots of observed TDS and simulated TDS by WQSAM for the same catchment.

5 Conclusions

The present study showed that the WQSAM model can be updated to take as input, monthly flow data generated by the WEAP model. This development dramatically increases the potential for application of the WQSAM model to catchments outside of South Africa, and offers the possibility of a water quality model suitable for semi-arid data-scarce catchments worldwide.

Data availability. The data used in this study are available at https://doi.org/10.6084/m9.figshare.5370799 (Slaughter and Mantel, 2017b).

Competing interests. The authors declare that they have no conflict of interest.

Special issue statement. This article is part of the special issue "Water quality and sediment transport issues in surface water". It is a result of the IAHS Scientific Assembly 2017, Port Elizabeth,

South Africa, 10–14 July 2017.

Edited by: Akhilendra B. Gupta
Reviewed by: Jagdish Kumar Bassin
and one anonymous referee

References

Adler, R. A., Claassen, M., Godfrey, L., and Turton, A.: Water, mining, and waste: an historical and economic perspective on conflict management in South Africa, The Economics of Peace and Security Journal, 2, 33–41, 2007.

Basson, M. S., Allen, R. B., Pegram, G. G. S., and van Rooyen, J. A.: Probabilistic Management of Water Resource and Hydropower Systems, Water Resources Publ., Colorado, USA, 424 pp., 1994.

Britton, D. L., Day, J. A., and Henshal-Howard, M.-P.: Hydrological response during storm events in a South African mountain catchment: the influence of antecedent conditions, Hydrobiologia, 250, 143–157, 1993.

Britz, T. J. and Sigge, G. O.: Quantitative investigation into the link between irrigation water quality and food safety: Volumes 1–4. Water Research Commission report No. 1773/1/12, Water Research Commission, Pretoria, South Africa, 2012.

Dimova, G., Tzanov, E., Ninov, P., Ribarova, I. and Kossida, M.: Complementary use of the WEAP model to underpin the development of SEEAW physical water use and supply tables, Procedia Engineering, 70, 563–572, 2014.

Gorgens, A. H. M. and de Clercq, W. P.: Research on Berg River Water Management – Summary of Water Quality Information System and Soil Quality Studies. Water Research Commission Report No TT252/06. Water Research Commission, Pretoria, South Africa, 2006.

Hamlat, A., Errih, M., and Guidoum, A.: Simulation of water resources management scenarios in western Algeria watersheds using WEAP model, Arab Journal of Geosciences, 6, 2225–2236, 2013.

Hollermann, B., Giertz, S., and Diekkruger, B.: Benin 2025-Balancing Future Water Availability and Demand Using the WEAP "Water Evaluation and Planning" System, Water Resour. Manag., 24, 3591–3613, 2010.

Hughes, D. A. (Ed.): SPATSIM, an Integrating Framework for Ecological Reserve Determination and Implementation. WRC Report No. TT 245/04. Water Research Commission, Pretoria, South Africa, 2004.

Hughes, D. A. and Slaughter, A. R.: Daily disaggregation of simulated monthly flows using different rainfall datasets in southern Africa, Journal of Hydrology: Regional Studies, 4, 153–171, 2015.

Hughes, D. A. and Slaughter, A. R.: Disaggregating the components of a monthly water resources system model to daily values for use with a water quality model, Environ. Modell. Softw., 80, 122–131, 2016.

Hughes, D. A., Hannart, P., and Watkins, D.: Continuous baseflow separation from time series of daily and monthly streamflow data, Water SA, 29, 43–48, 2003.

Levite, H., Sally, H., and Cour, J.: Testing water demand management scenarios in a water-stressed basin in South Africa: application of the WEAP model, Phys. Chem. Earth, 28, 779–786, 2003.

Li., X., Zhao, Y., Shi, C., Sha, J., Wang, Z.-L., and Wang, Y.: Application of Water Evaluation and Planning (WEAP) model for water resources management strategy estimation in coastal Binhai New Area, China, Ocean Coast. Manage., 106, 97–109, 2015.

Lynch, S. D.: Development of a raster database of annual, monthly and daily rainfall for Southern Africa. WRC Report No. 1156/1/04, Water Research Commission, Pretoria, South Africa, 2004.

Mallory, S. J. L., Odendaal, P., and Desai, A.: The Water Resources Modelling Platform User's Guide v3.3. Unpublished document available from stephen@waterresources.co.za, 2011.

Mantel, S. K., Hughes, D. A., and Slaughter, A. S.: Water Resources Management in the Context of Future Climate and Development Changes: A South African Case Study, Journal of Water and Climate Change, 6, 772–786, 2015.

Middleton, B. J. and Bailey, A. K.: Water Resources of South Africa, 2005 Study (WR2005), WRC Report No. TT381/08, Water Research Commission, Pretoria, South Africa, 2008.

Nash, J. E. and Sutcliffe, J. V.: River flow forecasting through conceptual models. Part I – a discussion of principles, J. Hydrol., 10, 282–290, 1970.

Pelletier, G. J., Chapra, S. C., and Tao, H.: QUAL2Kw – A framework for modeling water quality in streams and rivers using a genetic algorithm for calibration, Environ. Modell. Softw., 21, 419–425, 2006.

Piesold, K., Moahloli, G., Steyn, K., and Shand, N.: Berg River Dam: Further Water Quality Studies, Reservoir Modelling Report Report No. C201.51-171 R, 2007.

Pitman, W. V.: A mathematical model for generating river flows from meteorological data in South Africa. Report no. 2/73, Hydrological Research Unit, University of the Witwatersrand, Johannesburg, South Africa, 1973.

Schulze, R. E. (Ed.): ACRU: Background Concepts and Theory. ACRU Report No. 36, Dept. Agric. Eng., Univ. of Natal, Pietermaritzburg, RSA. SCHULZE RE (1995) Hydrology and Agrohydrology: A Text to Accompany the ACRU 3.00 Agrohydrological Modelling System, WRC, 1989.

Schulze, R. E. and Maharaj, M.: Development of a Database of Gridded Daily Temperatures for Southern Africa. WRC Report No. 1156/2/04. Water Research Commission, Pretoria, South Africa, 2004.

Sieber, J. and Purkey, D.: Water Evaluation and Planning System. User Guide for WEAP21. Stockholm Environment Institute, Massachusetts, USA, available at: http://www.weap21.org (last access: 1 September 2017), 2007.

Simate, G. S. and Ndlovu, S.: Acid mine drainage: challenges and opportunities, Journal of Environmental Chemistry and Engineering, 2, 1785–1803, 2014.

Slaughter, A. R.: Simulating microbial water quality in data-scarce catchments: an update of the WQSAM model to simulate the fate of *Escherichia coli*, Water Resour. Manag., 1–14, https://doi.org/10.1007/s11269-017-1743-1, 2017.

Slaughter, A. R. and Mantel, S. K.: The validation of algal growth processes in a water quality model using remote sensing data, Proceedings of the 8th International Environmental Modelling and Software Society (iEMSs) biennial meeting, Toulouse, France, 10–14 July 2016, http://scholarsarchive.byu.edu/cgi/

viewcontent.cgi?article=1374&context=iemssconference (last access: 1 September 2017), 2016.

Slaughter, A. R. and Mantel, S. K.: Land cover models to predict non-point nutrient inputs for selected biomes in South Africa, Water SA, 43, 499–508, 2017a.

Slaughter, A. R. and Mantel, S. K.: Water quality modelling of an impacted semi-arid catchment using flow data from the WEAP model: data and graphs, https://doi.org/10.6084/m9.figshare.5370799, 2017b.

Slaughter, A. R., Hughes, D. A., and Mantel, S. K.: The development of a Water Quality Systems Assessment Model (WQSAM) and its application to the Buffalo River catchment, Eastern Cape, South Africa, Proceedings of the 6th International Environmental Modelling and Software Society (iEMSs) biennial meeting, Leipzig, Germany 1–5 July 2012, ISBN: 978-88-9035-742-8, http://www.iemss.org/sites/iemss2012//proceedings/I2_2_0497_Slaughter_et_al.pdf (last access: 1 September 2017), 2012.

Slaughter, A. R., Retief, D. C. H., and Hughes, D. A.: A method to disaggregate monthly flows to daily using daily rainfall observations: model design and testing, Hydrological Sciences Journal, 4, 153–171, https://doi.org/10.1080/02626667.2014.993987, 2015.

Slaughter, A. R., Mantel, S. K., and Hughes, D. A.: Water Quality Management in the Context of Future Climate and Development Changes: A South African Case Study, Journal of Water and Climate Change, 8, jwc2016138, https://doi.org/10.2166/wcc.2016.138, 2016.

Slaughter, A. R., Hughes, D. A., Retief, D. C. H., and Mantel, S. K.: A management-oriented water quality model for data scarce catchments, Environ. Modell. Softw., 97, 93–111, 2017.

South African Department of Water Affairs (DWAF): Development of a Reconciliation Strategy for the Amatole Bulk Water Supply System. Final Report. Two Volumes. Prepared by SSI Engineers and Environmental Consultants on behalf of the Department of Water Affairs and Forestry Director, National Water Resource, Pretoria, South Africa, 2008.

Stirzaker, R., Biggs, H., and Roux, D.: Requisite simplicities to help negotiate complex problems, AMBIO, 39, 600–607, 2010.

Strahler, A. N.: Quantitative Analysis of Watershed Geomorphology, Trans. Am. Geophys. Union, 38, 913–920, 1957.

Verhoeven, J. T. A., Arheimer, B., Yin, C., and Hefting, M.: Regional and global concerns over wetlands and water quality, Trends Ecol. Evol., 21, 96–103, 2006.

Proc. IAHS, 377, 35–40, 2018
https://doi.org/10.5194/piahs-377-35-2018
© Author(s) 2018. This work is distributed under
the Creative Commons Attribution 3.0 License.

Open Access
PIAHS

Estimation of sediment deposits in the Ghézala reservoir in northern Tunisia

Majid Mathlouthi[1] **and Fethi Lebdi**[2]

[1]Research Laboratory in Sciences and Technology of Water at National Agronomic Institute of Tunisia, 43 avenue Charles Nicolle, 1082 Tunis, Tunisia
[2]National Agronomic Institute of Tunisia, 43 avenue Charles Nicolle, University of Carthage, 1082 Tunis, Tunisia

Correspondence: Majid Mathlouthi (majid_mathlouthi@yahoo.fr)

Received: 5 June 2017 – Revised: 11 December 2017 – Accepted: 14 December 2017 – Published: 16 April 2018

Abstract. The control of sedimentation in a reservoir provides a global evaluation of the process of erosion and transportation of sediment. Knowledge of sedimentation is useful for reservoir management. Bathymetric surveys can be used to assess the silting volume of dams. The results of two surveys of the Ghézala dam reservoir in northern Tunisia are available. The measurements provide initial information about the quantity and variability of silting and the mechanism of sediment deposition. According to the results of measurements, the average annual specific sediment yield of the Ghézala dam watershed is estimated at 1851 $\mathrm{t\,km^{-2}\,yr^{-1}}$. The annual average sediment volume trapped varies from 23 000 $\mathrm{m^3}$ in 1993 to 66 692 $\mathrm{m^3}$ in 2011. The sedimentation rates increases from 0.20 to 0.57 % overtime. The results indicate interdependence between the specific erosion rates and the occurrence of soils on steep slopes. The pressure exerted on the soil by plowing as well as overgrazing to meet the needs of the population of this area has exposed the soil to continued deterioration manifested by increased erosion endangering the only source of revenue for the area.

1 Introduction and objectives

Rainfall enhances sediment transport via watershed runoff, especially at the beginning of the rainy season. The stream flow detaches more river bed particles through floods as a result of the high flow velocity, which increases both the capacity (the total sediment mass) and the competence (the maximum particle size) that a stream can carry. When a stream enters a reservoir, the flow velocity and energy decrease, causing the sediment to be deposited. The coarse sediment is typically deposited at the reservoir mouth, while the fine particles are deposited along the water flow direction in the reservoir. In Tunisia, 45 % of the land area is threatened by erosion (Boussema, 1996). On average 25.3 million $\mathrm{m^3}$ of sediment is deposited annually in dammed reservoirs. As noted above, a major problem associated with dams is sediment deposition, which increases flood risk because of a decreased water storage capacity, with attendant effects on the flood water storage capacity.

The reduced storage capacity affects the safe yield of the reservoir to meet different water demands. Various models have been suggested to study reservoir siltation. Campos (2001), for example, presented a three-dimensional model for flow and sediment concentration in reservoirs. He used two bed load functions, each one for bed direction. Laboratory test data and actual field data for Itaipu Reservoir in Brazil were considered for model verification. The result of laboratory tests indicated a good agreement between flow and sediment routing simulations and predictions of sediment deposition through the reservoir. Pak et al. (2010) evaluated a compatible reservoir sedimentation model with HEC-HMS, which was applied to two reservoirs to compare the model outputs with measured data. The results indicated a reasonable prediction of both daily flow rates and sediment loads. Issa et al. (2013) estimated the useful life of Mosul Dam, Iraq, based on a bathymetric survey approach. Mohammad et al. (2016) focused on studying the sediment deposition within Mosul Dam reservoir for the period from 1986

Published by Copernicus Publications on behalf of the International Association of Hydrological Sciences.

to 2011, using the HEC-RAS 4.1 model. Sediment depositional simulation was considered on monthly intervals (time step) for the period of study, while the pattern was considered for 5 year intervals (1991, 1996, 2001, 2006 and 2011). Thus, the simulation period extended from 1986 (beginning of the Mosul Dam operation) to 2011. Banasik et al. (2012) estimated the annual sediment yield of a small (91 km^2) agricultural catchment in central Poland. In this investigation, the suspended sediment input to a reservoir was estimated using the Universal Soil Loss Equation coupled with a sediment delivery ratio (USLE-SDR). The annual bed load was estimated based on the flow duration curve and three different bed load formulae. Reservoir bathymetry surveys were carried out four times. A close agreement was found between the amount of sediment deposited in the reservoir and the sediment input estimated using the USLE-SDR and the bed load formulae.

Kondolf et al. (2014) summarize collective experience from five continents in managing reservoir sediments and mitigating downstream sediment starvation. Drawdown flushing works best in narrow reservoirs with steep longitudinal gradients and with flow velocities maintained above the threshold to transport sediment. Walling and Webb (1996) provide an overview of knowledge of global sediment yields and identify outstanding uncertainties and research needs. They point out that the analysis of available longer-term records provides an important means of assessing the sensitivity of sediment yields to environmental change, which in turn requires consideration of the impact of both human activity and climate change. Walling et al. (2013) studied the changing fluvial sediment inputs to the world's deltas. The sediment loads of the world's rivers are highly sensitive to both human impact and climate change, and the sediment loads of many rivers have changed markedly in recent decades.

The aim of this article is to present the state of the alluvial deposition of a dam in Tunisia according to the available data and to study the evolution of its capacity over time. Among other things, efforts will be made to evaluate the effects of anthropogenic or climatic factors in the catchment area on alluvial deposition in the reservoir. Due to time constraints, it is not possible to detail the erosion mechanism. A method, derived from the assessment of sediment retention in Ghézala dam, is used in this study. Records from bathymetric surveys carried out in 1993 and 2011 are available and relatively reliable erosion rates can be calculated using the retained sediments in the reservoir.

2 Study area and dam characteristics

Ghézala dam is a small dam on the basin of Ichkeul Lake in Northern Tunisia (Fig. 1). It is a multipurpose dam for flood control, irrigation and water supply. The total water storage capacity of the Ghézala dam reservoir is 11.7×10^6 m^3 at a normal operation level of 82.5 m a.s.l. of which 9.1 and 2.6×10^6 m^3 are live and dead storages, respectively. The impoundment of the dam was in 1985. Sediment deposition is one of the major problems of dams, such as Ghézala dam, particularly in the upper part of the reservoir. Its watershed covers an area of 48 km^2 (Fig. 1). Its lithology is highly varied with limestone, clay and marl. The most representative soils are different types of mineral soils, calcimorphic soils (rendzina, brown calcareous soils) and brown soils with sparse vegetation cover. The climate is subhumid. The mean annual precipitation is 680 mm (Mathlouthi and Lebdi, 2010) and the minimum and maximum monthly mean air temperatures are 13 and 23 °C, respectively.

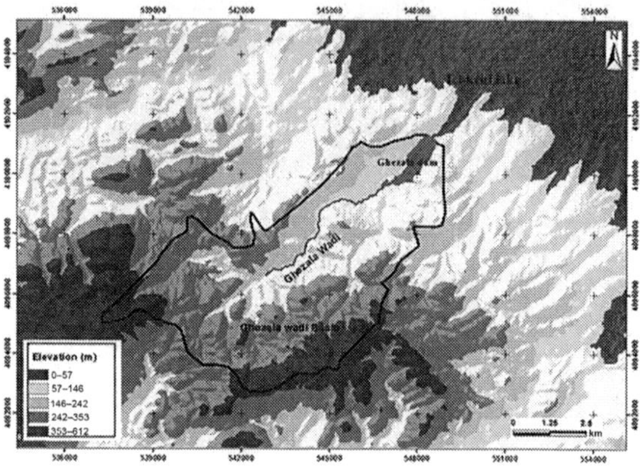

Figure 1. Location of Ghézala dam.

3 Methodology

Measurement of siltation in reservoirs normally involves a bathymetric survey of the flooded portion of the reservoir and a direct topographical survey of the non-flooded portion between the normal water level and the level of water during the measurement. The bathymetric survey is based on a simple comparison of reservoir morphology at two different periods of time, first at the time of the construction of the dam and second at the time of the survey. In this study bathymetric survey of the reservoir bed was conducted along transverse profiles previously identified, the depths being recorded by a scrolling Echo Sounder. This method is uncomplicated and easy to use, but it is not very precise and cannot be used systematically for depths less than 5 m (Claude and Chartier, 1977). For this study, the choice of the cross section on the reservoir to be surveyed is made on a topographical plan of the region of the dam reservoir at the scale 1 : 5000 and with contour spacing at 2 m. There is no specific rule to guide this choice; we generally seek cross sections to the bed of the wadi, avoiding the narrowing and preferably choosing the widest sections. If the reservoir is divided into several branches, each one is treated indepen-

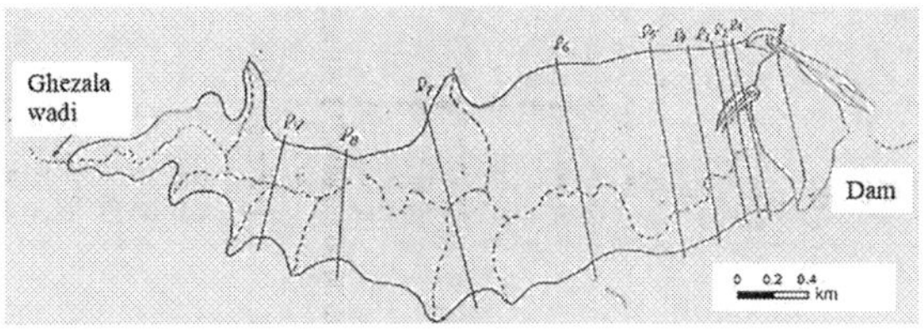

Figure 2. Location of the sediment profiles surveyed in the Ghézala dam.

dently. Nine cross sections of the reservoir of Ghézala wadi were identified (Fig. 2). For each profile, the original (initial) cross section of the reservoir at the time of impoundment in 1985 was obtained from the topographical plan. Then bathymetric surveys were conducted to establish the current reservoir depths. The new profile is compared with the original profile. The two profiles thus obtained and the silted surface of the profile is quantitatively determined in a GIS, in other words the area of siltation (in m^2) of a section at the height of the water site to which the measurements were made.

In this study the General Method was used to determine the volume of siltation. This method is based on silted sections of the surveyed profiles but also takes into account the angles between adjacent sections, shown as P_a and P_b in Fig. 3 (Claude and Chartier, 1977; Alahiane et al., 2014).

First the sediment surfaces, S_a and S_b respectively, are determined of the two retaining profiles P_a and P_b and then the distance between these two profiles is calculated (see Fig. 3). To do this we draw a line ab between the centre points of P_a and P_b; we draw the line cd perpendicular to segment ab at its center; the distance between the middle (e) of segment cd and points a and b give the lengths l_a and l_b.

Determing the total volume takes into account the slopes αa and αb, respectively for P_a and P_b compared to straight sections ea and eb. The total volume V_{ab} is (Claude and Chartier, 1977):

$$V_{ab} = S_a \sin\alpha a l_a + S_b \sin\alpha b l_b \quad (1)$$

Where αa is the angle $\widehat{eaP_a}$ and αb is the angle $\widehat{ebP_b}$.

4 Results and discussion

Since the impoundment of the dam (1985), two measures of siltation of the Ghézala dam were carried out by the General Directorate of Dams and Large Hydraulic Works (Tunisia) during relatively calm weather (light wind). The results of these measurements are shown in Table 1.

Erosion rates were estimated by the sediment volume measured in the reservoir divided by the area of the catchment draining into the reservoir. At the time of the first measurement, the amount of silt was accumulated for 8 years and 5 months. The reservoir is silted at 1.64 % of its capacity, which allows us to assess the annual rate siltation at

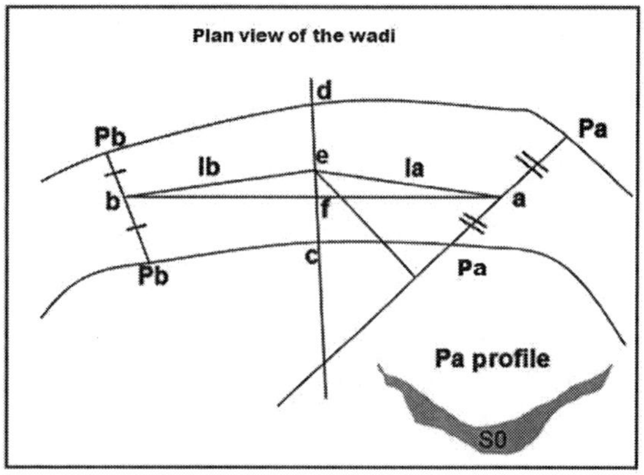

Legend: S0 = silted section on the survey profile Pa

Figure 3. Diagram illustrating the General Method for estimating sedimentation volume. A plan view is shown in the upper part and a cross-section profile view is shown in the bottom right hand corner.

23 000 $m^3\,yr^{-1}$ or 0.25 % of the live capacity of the reservoir (the live capacity of the reservoir is $9.1 \times 10^6\,m^3$). In the second measurement, 14.8 % of the reservoir capacity has been infilled, corresponding to an annual siltation rate of 66 692 $m^3\,yr^{-1}$ which gives a life expectancy of about 175 years before the reservoir has been totally infilled. The annual rate of siltation has increased (threefold from 1993 to 2011) which reduces the life of the dam from about 500 to 175 years. This can be explained by clearing of maquis vegetation in the watershed due to an exponential expansion of annual crops. The latter activity involves plowing in the direction of the slope, a factor that accentuates and exacerbates erosion in the watershed. This is evident in the estimated erosion from the watershed which increased from 0.48 to 1.39 $mm\,yr^{-1}$ during the same time period.

In Fig. 4 we compare the profile along the Ghézala river before impoundment with the measured profiles along the silted bottom (low point of the current sections). Figure 5 shows the cross sectional areas of silted sections measured along the profiles in Ghézala reservoir in May 1993. We note here the importance of siltation in the downstream (see Fig. 2) part of the reservoir.

Ghézala dam has been operational since April 1985; the latest measurements therefore include a period of 26 years of operation. The dam operating results are summarized in Table 2. To calculate the total sediment input to the dam we assumed that the volume of water released had a mean load of 3 $g\,L^{-1}$ (or $kg\,m^{-3}$) of sediment (Ben Mammou and Louati, 2007), assuming the same value as for the Mellègue dam calculated over 14 years. This assumption depends on the morphology, the pedology, the watershed land use and rainfall patterns. Further, the spilled water was assumed to contain 1 $g\,L^{-1}$ of sediment taking into account the morphology of the reservoir, based on spot measurements made during flood

Table 1. Measurements of siltation in the Ghézala dam reservoir.

Measurement date	Water level (m)	Siltation (10^6 m^3)	Annual siltation (m^3)	Reservoir capacity infilled with sediment (%)	Reservoir capacity infilled with sediment per year (%)	Erosion of the watershed (mm yr^{-1})
May 1993	82.3	0.192	23 000	1.64	0.20	0.48
June 2011	81.3	1.734	66 692	14.8	0.57	1.39

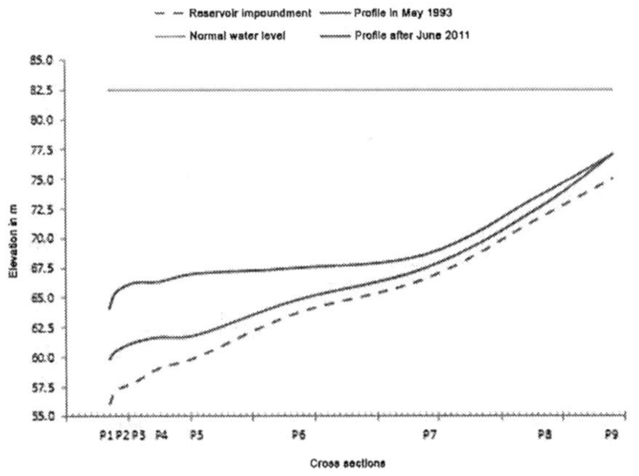

Figure 4. Profiles along Ghézala wadi before impoundment and after measurements.

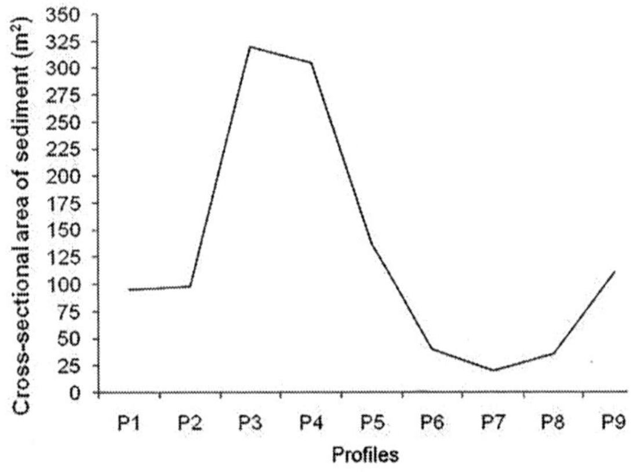

Figure 5. Cross-sectional areas of sediment at the measured profiles along the Ghézala river (in May 1993).

conditions. It is assumed that the solids density of sediments in the bottom of the reservoir comprised of silts and clay is 1.2 t m^{-3}, with a real density of 2.4 t m^{-3} and a bulk density of 1.7 t m^{-3}.

From these assumptions the balance of sediment transport in 2011 may be estimated (Table 3).

The volumes of water flowing into the reservoir or withdrawn are well controlled. The volumes that are discharged

Table 2. Operation of Ghézala dam during period 1985–2011 (48 km^2).

	Inflows (10^6 m^3)	Spilled water (10^6 m^3)	Release (10^6 m^3)
Total over 26 years	232.49	77.029	50.906
Annual average	8.942	2.963	1.958

by the spillway during extreme floods are controlled, however their turbidity is unknown. It is estimated, however, that floodwaters released from the reservoir contain little more than 3 g L^{-1} due to sediment deposition before the arrival at the spillway. The bulk density and true density of sediment deposited and drawdown silt are poorly understood, consolidated sediment volumes were measured for several years under several metres of water, and an average concentration of solids (g L^{-1}) was measured on the liquid flow. The transition from one to the other requires a precise knowledge of the densities; which is not easy to achieve. At the watershed scale of 48 km^2, the transport of 1.925×10^6 m^3 of silt equates to 2.31×10^6 t in 26 years, representing an average erosion rate from the watershed of 1542 m^3 km^{-2} yr^{-1} or 1851 t km^{-2} yr^{-1} resulting in an average concentration of the water entering the reservoir of 10 g L^{-1} solids.

Based on the results of a large 2 year intensive measurement campaign of the turbidity of the main river in the extreme north of Tunisia and watersheds of Lake Ichkeul, the annual average turbidity of Ghézala river is assumed to be 6.56 g L^{-1}, which equates to 1.547×10^6 t or 1.289×10^6 m^3 contributions of solids. The difference between the measured siltation in the reservoir (neglecting the other two values, turbidity of released and spilled water, not measured) can be attributed to long distance transportation and therefore represent thrusting 0.445×10^6 m^3 (1.289×10^6 m^3 compared to 1.734×10^6 m^3) or 25 % of total sediment transport (1.734×10^6 m^3) and 34 % of transportation suspended (1.289×10^6 m^3). The reservoirs of Tunisian dams lose 0.5 to 1 % of their capacity each year due to infill with alluvium. The annual average sediment volumes trapped in Tunisian dam reservoirs varies from 0.017×10^6 m^3 for Bezirk dam (Northeast Tunisia) to 6×10^6 m^3 for Sidi Salem dam, North Tunisia. Siltation measured in 2002 in Sejenane dam located in the same region as the Ghezala dam resulted in an es-

Table 3. Results of sediment transport in Ghézala dam during 1985–2011.

	Measured siltation	Sediment in released water	Sediment in spilled water	Total sediment yield
Volume (10^6 m^3)	1.734	0.127	0.064	1.925

Table 4. Increase in agricultural area over time (based on interpretation of aerial photographs and satellite images in both periods).

Year	Land use as % of the total area of the Ghézala dam catchment (48 km^2).			
	Herbs and shrubs	bare lands	arable land	Fruit trees
1979	39.8	40	20.2	–
2011	18.7	2.03	74.6	4.67

timated average annual siltation rate of 0.5×10^6 m^3 (Ben Mammou and Louati, 2007).

Field observation shows that the watershed is subject to intense erosion where mass movements and gullies predominate, accentuated by plowing and soil preparation for cultivation. The vulnerability of soil in the watershed to erosion is attributed to lithology. The dominant formations (Upper Cretaceous) consists mainly beds of clays and marls with very little resistance to erosion (clays and marls) and harder chalky beds. In the Ghézala dam watershed the steepest slopes correspond to limestone geology. Several authors have shown that specific erosion increases significantly with slope (Huang, 1995; Jaoued et al., 2005). The authors attribute such a relationship to the nature of the soil, the speed of flow, the characteristics of rainfall and surface crusting (Jaoued et al., 2005).

The combined action of various factors of the natural environment and increased human agricultural activities has contributed to the degradation of soils in the Ghézala watershed (Table 4). The pressure exerted on the soil by plowing as well as overgrazing to meet the needs of the population of this area has exposed the soil to continued deterioration manifested by different erosive phenomena endangering the only source of revenue for the area. The observed increase in sedimentation in the reservoir (in 2011 compared to 1993) could be caused by other factors, such as increased intensity and amount of rainfall during the study period. Figure 6 shows that the number of floods after 1993, the date of the first siltation measure, is important. Knowing that a hydrological year begins in September and ends in August of the succeeding year. Annual inflows exceeded twice the reservoir capacity, 23 and 24 million of m^3 respectively in 2003 and 2005. This increase in inflows between the time of the first and second sedimentation measurements and also withdrawal that is not performed at optimal times, otherwise the opening of the bottom outlet valves during extreme floods, have contributed to increasing the reservoir sedimentation rate. Additional questions may arise about the efficiency of the withdrawal, how and at what times can be conducted? With what flow? And for how long (throughout and after the floods)?

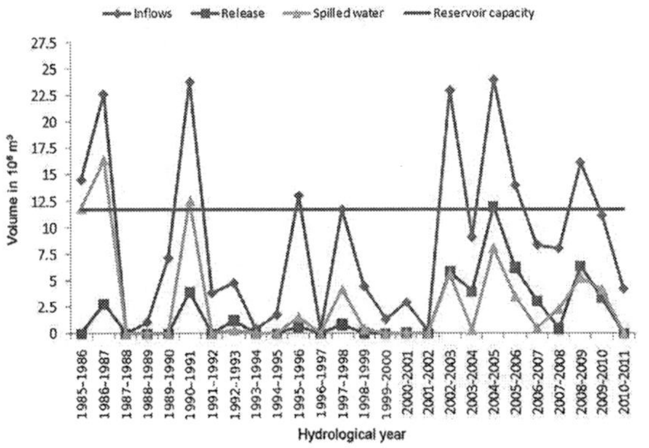

Figure 6. Annual inflows, release and spilled water at Ghézala dam (1985–2011).

5 Conclusions

Water erosion is a natural process which is considered to be the primary cause of degradation of agricultural land. It is necessary to take corrective action before soils become infertile. Measurements of siltation of Ghezala reservoir presented here, enable estimation of the volume of sediment trapped in the reservoir. It is suggested that the sediment trapped could be returned to agricultural land. The accuracy of the sedimentation estimation method depends upon the profiles selected and precise mapping of the reservoir bed. Siltation measures should be developed to monitor, at regular intervals, the evolution of the life time of dams and improve operating conditions. The results of these measures have highlighted the significant erosion to which the watershed of the Ghézala dam is subjected. The use across the watershed of a runoff and erosion model with distributed physical parameters can improve understanding of the variability of the factors responsible for soil loss and thus help to address the problems related to the soil conservation. Indeed, according to field surveys, the most significant factors for erosion across this study area are lithology, slope and land use.

The control of sedimentation in the reservoirs provides an integrated evaluation of the process of erosion and transportation of sediments taking account of the bed-material load. The knowledge of sedimentation is also useful for dam management. In Tunisia, little was known about erosion, transportation and sedimentation phenomenon which is necessary in order to manage their consequences.

Data availability. The data are printed in paper documents stored (archived) in office of the General Directorate of Dams and Large Hydraulic Works of the Ministry of Agriculture of Tunisia (http://www.agriculture.tn/). These data are the property of this organization responsible for dams in Tunisia, and are available in situ.

Competing interests. The authors declare that they have no conflict of interest.

Special issue statement. This article is part of the special issue "Water quality and sediment transport issues in surface water". It is a result of the IAHS Scientific Assembly 2017, Port Elizabeth, South Africa, 10–14 July 2017.

Acknowledgements. The authors acknowledge the help of General Directorate of Dams and Large Hydraulic Works of the Ministry of Agriculture of Tunisia.

The authors thank the reviewers for their relevant remarks that contributed to the improvement of this article.

Edited by: Kate Heal
Reviewed by: two anonymous referees

References

Alahiane, N., El Mouden, A., Lhaj, A. A., and Boutaleb, S.: Practical Method Proposed to Estimate Silting's Rate in Small and Hillside Dams, J. Water Resour. Protect., 6, 930–943, 2014.

Banasik, K., Gorski, D., Popek, Z., and Hejduk, L.: Estimating the annual sediment yield of a small agricultural catchment in central Poland, in: Erosion and Sediment Yields in the Changing Environment, edited by: Collins, A. L., Golosov, V., Horowitz, A. J., Lu, X., Stone, M., Walling, D. E., and Zhang, X., IAHS Publ., 356, 267–275, 2012.

Ben Mammou, A. and Louati, M. H.: Évolution temporelle de l'envasement des retenues de barrages de Tunisie, Revue des sciences de l'eau, 20, 201–210, 2007.

Boussema, M. R.: Système d'information pour la conservation et la gestion des ressources naturelles: in colloque international sur le rôle des technologies de télécommunications et de l'information en matière de protection de l'environnement, Tunis, Tunisia, 17–19 April 1996.

Campos, R.: Three-Dimensional Reservoir Sedimentation Model, PhD thesis, University of Newcastle, Newcastle, UK, 2001.

Claude, J. and Chartier, R.: Mesure de l'envasement dans les retenues de six barrages en Tunisie. Campagne 1975, Cah. ORSTOM, sér. Hydrol., 16, 3–35, 1977.

Huang, C.: Empirical analysis of slope and runoff for sediment delivery from interill areas, Soil Sci. Soc. Am. J., 59, 982–990, 1995.

Issa, I., Al-Ansari, N., and Knutsson, S.: Sedimentation processes and useful life of Mosul Dam reservoir, Iraq, J. Sci. Res. Eng., 5, 79–84, 2013.

Jaoued, M., Gueddari, M., and Saadaoui, M.: Modélisation de l'érosion hydrique dans le bassin versant de l'oued M'Khachbia (Nord-Ouest de la Tunisie), Geo-Eco-Trop., 29, 15–24, 2005.

Kondolf, G., Gao, Y., Annandale, G., Morris, G., Jiang, E., Zhang, J., Cao, Y., Carling, P., Fu, K., Guo, Q., Hotchkiss, R., Peteuil, C., Sumi, T., Wang, H., Wang, Z., Wei, Z., Wu, B., Wu, C., and Yang, C.: Sustainable sediment management in reservoirs and regulated rivers: Experiences from five continents, Earths Future, 2, 256–280, 2014.

Mathlouthi, M. and Lebdi, F.: Modélisation de la relation pluie–ruissellement par durée d'épisode pluvieux dans un bassin du Nord de la Tunisie, Hydrol. Sci. J., 55, 1111–1122, 2010.

Mohammad, E. M., Al-Ansari, N. A., Issa, I., and Knutsson, S.: Sediment in Mosul Dam reservoir using the HEC-RAS model, Lakes and Reservoirs: Research and Management, 21, 235–244, 2016.

Pak, J., Fleming, M., and Ely, P.: Assessment of reservoir trap efficiency methods using the hydrologic modeling system (HEC-HMS) For the upper north Bosque River watershed in central Texas, in: Proceedings of the 2nd Joint Federal Interagency Conference (9th Federal Interagency Sedimentation Conference and 4th Federal Interagency Hydrologic Modeling Conference), Las Vegas, Nevada, 27 June–1 July 2010.

Walling, D. and Webb, B.: Erosion and sediment yield: A global overview, in: Erosion and Sediment Yield: Global and Regional Perspectives: IAHS Publ., 236, 3–19, 1996.

Walling, D., Young, G., and Perillo, G.: Changing fluvial sediment inputs to the world's deltas, Deltas: Landforms, Ecosystems and Human Activities, 358, 12–25, 2013.

Proc. IAHS, 377, 41–50, 2018
https://doi.org/10.5194/piahs-377-41-2018
© Author(s) 2018. This work is distributed under
the Creative Commons Attribution 4.0 License.

Morphodynamic change analysis of bedforms in the Lower Orinoco River, Venezuela

Santiago Paul Yepez[1], Alain Laraque[1], Carlo Gualtieri[2], Frédéric Christophoul[1], Claudio Marchan[3], Bartolo Castellanos[4], Jose Manuel Azocar[5], Jose Luis Lopez[4], and Juan Alfonso[5]

[1]GET, UMR CNRS/IRD/UPS – UMR 5563 du CNRS, UMR234 de l'IRD, 31400 Toulouse, France
[2]University of Napoli Federico II, 80138 Napoli, Italy
[3]Centro de Sismología, Universidad de Oriente 6101, Venezuela
[4]IMF – UCV Instituto de Mecánica de Fluidos, Caracas 1041-A, Venezuela
[5]COEA, Instituto Venezolano de Investigaciones Científicas (IVIC), Caracas 1020-A, Venezuela

Correspondence: Santiago Paul Yepez (syepez14@gmail.com)

Received: 27 June 2017 – Revised: 15 December 2017 – Accepted: 11 January 2018 – Published: 16 April 2018

Abstract. The Orinoco River has the third largest discharge in the world, with an annual mean flow of $37\,600\,m^3\,s^{-1}$ at its outlet to the Atlantic Ocean. Due to the presence of the Guiana Shield on the right bank, the lower reach of the Orinoco has a plan form characterized by contraction and expansion zones. Typical 1–1.5 km wide narrow reaches are followed by 7–8 km wide reaches. A complex pattern of bed aggradation and degradation processes takes place during the annual hydrological regime. A series of Acoustic Doppler Current Profiler (ADCP) transects were collected on an expansion channel in the Orinoco River, specifically over a fluvial island, representative of the lower Orinoco. In this study, temporal series of bathymetric cartography obtained by ADCP profiles combined with Differential Global Position System (DGPS) measurements (with dual-frequency), were used to recover the local displacement of bed forms in this island. The principal aims of this analysis were: (1) to understand the dynamics and evolution of sand waves and bars at this section and (2) to quantify the volume (erosion vs. accretion) of a mid-channel bar with dunes by applying DEM of Difference (DoD) maps on time series of bathymetric data. This required sampling with ADCP transects during the months of: May 2016; November 2016 and April 2017. Each bathymetric transect was measured twice, 1 day apart and on the same trajectory obtained by a GPS receptor. The spatial analysis of these ADCP transects is presented as a novel tool in the acquisition of time series of bathymetry for a relatively deep section (~ 20 m) and under variable flow conditions.

1 Introduction

The processes of erosion and sediment transport are key components for quantifying the formation of mid-channel bars. Mat Salleh and Ariffin (2013), categorized the fluvial river system into three zones: (1) an erosion zone of runoff production and sediment sources, (2) a transport zone of water and conveyance and (3) a deposition zone of runoff delivery and sedimentation.

River banks have been shown to act as key sediment sources in many different drainage basins, and can supply over 50 % of catchment sediment output (Lawler et al., 1999).

The volumetric change associated with bedload flux difference and its quantification is one of the main objectives of many investigations to understand sediment transport in rivers. However, in many cases this type of analysis is performed in a two-dimensional way. An example of this is the delineation of the riverbank from remotely sensed imagery to calculate accretion and erosion areas, using both airborne platforms and satellite platforms. In the last decades, developments in Remote Sensing and GIS technologies have allowed geomorphologists to develop campaigns of acquisition of Digital Elevation Models (DEMs) at much high precision and temporal frequency to analyze landform evolution

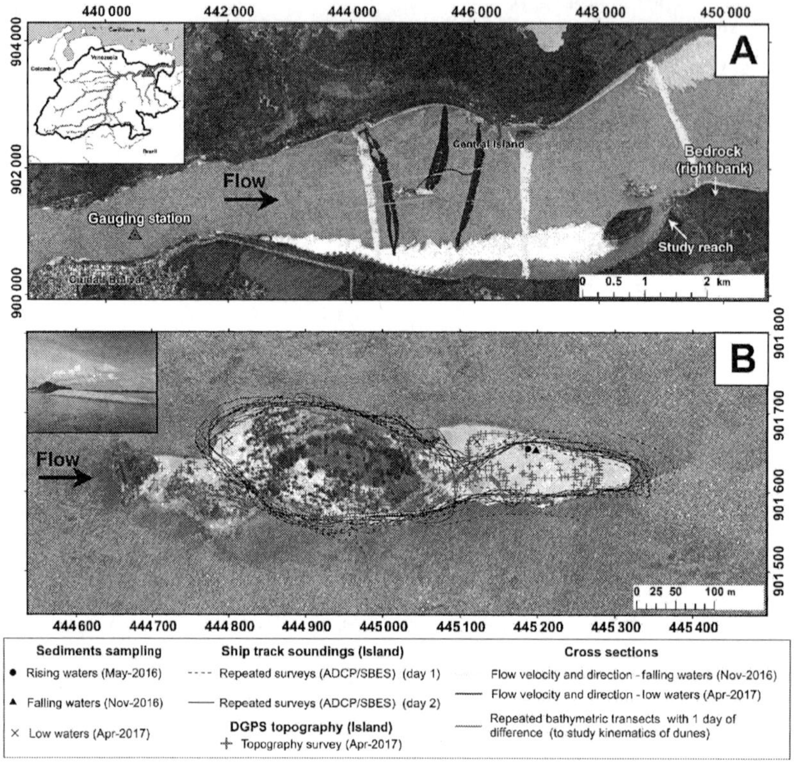

Figure 1. **(a)** shows ~ 7 km of the study site (Lower Orinoco) drawn with the red dotted line. Channel pattern is developed in a zone of expansion downstream from the gauging station. **(b)** illustrates the trajectories of the ADCP repeated transects performed during the field campaign.

and the change (Fuller et al., 2002; Williams et al., 2011; Williams, 2012; Carrivick et al., 2013).

In this work, a combined survey strategy using ADCP, Single Beam Echo-sounder (SBES) and DGPS equipments was implemented. The ADCP, mounted on a moving boat, collected data for mapping the flow field, and the SBES, on the same vessel, surveyed the channel bathymetry and the bed forms that create friction losses. Currently, the study of the processes linked to sediment transport, flow dynamics and erosion is a complex task, due in part to the technical and logistical constraints of deploying accurate data acquisition equipment to characterize morphological changes with recent data (Leyland et al., 2016).

The purpose of this research is to describe in a study reach of the lower Orinoco, the methods and advantages of repeated surveys by an ADCP and bathymetric soundings within a mid-channel bar affected by fluvial erosion/deposition. The specific research aims are: (1) to understand the mechanics of formation and evolution of sand waves and bars at this river section and (2) to quantify the volume (erosion vs. deposition) of a mid-channel bar, applying DoD maps (change detection) based on time series of bathymetric and topographic data. Comparisons between state of the art and traditional techniques are made to evaluate the effectiveness of these methodologies.

2 Study area

The Orinoco Basin is located (Fig. 1) in the north of South America between 2 and 10° N and 75 and 61° W. 70 % of this basin lies in Venezuela, while the remainder 30 % covers part of Colombia (Silva León, 2005). The headwaters of the Orinoco are at 1047 m above sea level and flow from its source at the Cerro Delgado–Chalbaud, in the Parima Range. The main channel covers a distance of 2140 km (Silva León, 2005) from the Brazilian border to its mouth in the Atlantic Ocean.

The basin comprises three large geographic zones: (1) the Andes and Caribbean Coastal Ranges, where most suspended sediments originate; (2) the Llanos (floodplain area), which are crossed by all major tributaries coming from the Andes and (3) the Precambrian Guiana Shield from which drains predominantly black water with very low suspended sediment content (Lewis Jr. and Saunders III, 1984, 1989;

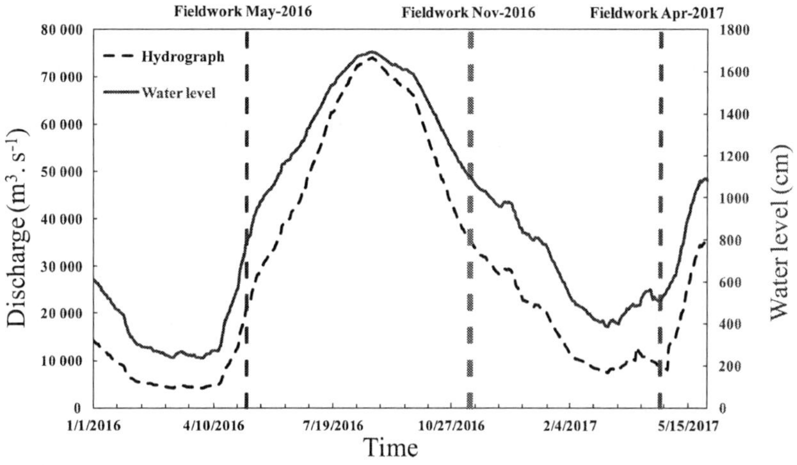

Figure 2. Daily recorded discharge and gauge data from Ciudad Bolivar station during the study period (2016–2017).

Paolini et al., 1987; Depetris and Paolini, 1991; López and Perez-Hernandez, 1999; Warne et al., 2002; Laraque et al., 2013; Yepez et al., 2016).

The study reach is characterized by numerous sand bars and islands that divide the flow into several channels (López and Perez-Hernandez, 1999).

The annual hydrological regime of the lower Orinoco at its main flow gauging station at Ciudad Bolivar in Venezuela is characterized by a unimodal regime where the rising waters begin in April, and maximum streamflow is reached in August (Fig. 2). The amplitude between the low and high waters is ~ 16 m.

For this study, a mid-channel bar (~ 700 m length) representative of many of the banks and flanks of the lower Orinoco, where erosion and accretion of sediments occurs was selected as the study site.

This large mid-river fluvial island is entirely separated from the floodplain by water, exhibits stability, is flooded during bankfull conditions, and has established permanent vegetation. This sector is associated with an open channel characterized by alternating contraction and expansion zones. It is located at a distance of 4 km downstream from the main gauging station of Ciudad Bolivar (Fig. 1). The flanks or banks of the mid-channel bar (Central island) at the study site (Fig. 3) are between 1–10 m in height, although locally it can reach 13 m. The banks are composed of highly erodible and poorly consolidated sand deposits ($D_{50} = 369.55$ µm).

Figure 3. The banks of the mid-channel bar (study area) are associated with highly erodible unconsolidated deposits.

3 Data and methods

3.1 Bathymetric and topographic data

Equipment such as ADCP are essential tools to measure the flow discharge in large rivers, especially when these rivers are dominated by unstable flows. The data associated with the velocity vector components obtained from ADCP equipment can be accurately plotted if this equipment is combined with dual frequency GPS receivers.

Time-based simulations were built to understand the sediment dynamics and flow in this sector. These simulations were generated from ADCP data sequences and bathymetric profiles combined with DGPS measurements (Table 1). The combination of both data allowed study of the flow dynamics related to river bank erosion/deposition over a fluvial island in the lower Orinoco.

Using the methodology implemented by Dinehart and Burau (2005), bathymetry was mapped on the mid-channel bar with ADCP equipment. The distances were measured along four transducer beams to the bed of the river. Subsequently, each of these distances is internally converted to a depth. The four depths obtained by each beam are recorded and averaged by the ADCP to provide one measurement between the device and the bed. However, each of these distances is suffi-

Table 1. SBES and ADCP surveys acquired on three occasions in the Orinoco River.

Survey campaign	Start	Finish	Number of surveys	water level in cm	Mean discharge in $m^3 s^{-1}$
			(SBES)		
Rising waters 2016	07:20, 7 May	12:12, 7 May	9	783	20 665
	07:05, 8 May	12:05, 8 May	9	800	20 994
			(ADCP + SBES)		
Falling waters 2016	11:00, 11 Nov	17:00, 11 Nov	9	1101	35 866
	08:00, 12 Nov	14:00, 12 Nov	10	1098	35 546
	08:00, 13 Nov	14:00, 13 Nov	3	1091	35 016
			(ADCP)		
Low waters 2017	10:00, 21 Apr	17:00, 21 Apr	14	515	7119
	08:00, 22 Apr	17:00, 22 Apr	16	518	9538

ciently accurate to create a precise bathymetry at relatively shallow depths. The orientation of each of the four transducer beams is at 20° downward from the vertical. Two of the beams are aligned at 20° from the vertical to port and starboard, while the two remaining beams maintain the same angle forward and backward. The cone angle of each beam is 2°, which ensures a very small footprint between the beam contact and the bed. The coordinates are rotated using the pitch and roll measurements performed by ADCP, which allows the true coordinates to be obtained.

For the acquisition of topographic data, RTK DGPS (Real Time Kinematic – Differential GPS) equipment was used, which in dynamic mode allows collection of points of coordinates and altitude every second. The DGPS system allows collection of topographic data with a high spatial and temporal resolution, and altitude of great accuracy. Thus, these data can be used to evaluate the processes and changes associated with the fluvial erosion of river banks and islands.

The VMT (Velocity Mapping Toolbox) software (Parsons et al., 2013) allows extraction of the cloud of points in depth (meters) from the ADCP bathymetry transects. The repeated transects used for bathymetric surveying and the topographic survey on the fluvial island was realized with two devices: (1) a River Ray 600 KHz RDI Teledyne coupled to a Hemisphere Crescent VS100 series GPS compass (in differential mode) and (2) two GPS receivers RTK S9 (dual frequency) brand Stonex (cm-level accuracy) using the WGS84 System. Additionally, a Single Beam Echo-sounder (SBES) GPSmap 62S GARMIN was used to compare measurements with repeated transects from ADCP measurements around the fluvial island. Since a known control point did not exist in the study area, it was decided to place a fixed surveying station in the middle of the island, whose position is referred to the SIRGAS-REGVEN National Geodetic System. The measurement phase was performed using the static differential mode, with a duration of 4 h, 5 s for capture intervals and 10° of cut off elevation. For the calculation, analysis and processing of the differential data the AUSPOS Online service mode was used. A total of 371 points (topography) were measured from 1 to 10 s in RTK mode. The baseline did not exceed 552 m. To obtain the orthometric heights from the ellipsoidal heights, the Venezuelan Geoidal Model 2017 – VGM17 (Acuña, 2017) was used.

In this study, the data of bathymetric transects from 2016 to 2017 were analyzed. Each series of the ADCP transects has between 69 100 points (corresponding to 8.41 ha) and 131 963 points (corresponding to 16.24 ha). A first field campaign around the island was conducted out in high waters, and later the second one was carried out in the low water stage. The differences in the number of points surveyed in the two field campaigns is mainly due to the greater exposure of the fluvial island during the low water stage in April. For the construction of the digital elevation models repeated transects were surveyed around the island (Fig. 1). By combining topography and bathymetry information at the same time, it is possible to create a single model of the interface between the river bed, the submerged slope of the island and the exposed part of the island.

This analysis was based on the methodologies of Wheaton et al. (2010), which focus on detecting and quantifying surface and volumetric changes in rivers. For our study these methodologies were applied to different stages of the annual hydrological regime. Terrain digital elevation models were created from data processing with the ESRI ArcGIS program. In this step, our analysis was based on a workflow where survey points were used to perform a Triangular Irregular Network (TIN) applying Delaunay triangulation. Subsequently, the TIN model was linearly resampled in a grid with a specific resolution defined by the user. A polygon feature was built around the cloud of points and then used as a mask in the construction of the TIN model to mitigate the effects of the interpolation beyond the area where the points were acquired. A resolution of 1 m was selected for both DEMs used in this analysis, as a balance between computational time and loss of information. Thus, we have been able to define a resolution fine enough to resolve the morphological changes at the scale of islands and bars. A DoD was calculated from the subtraction of the elevations of each DEM generated on a different date on a cell-by-cell basis. The net volumetric change estimate was calculated by multiplying the elevation change (measured as depth in m) by the surface area of each cell (1 m^2 for this case). Finally, these volumes were added to erosion and deposition categories to obtain the net volumetric budget.

An error evaluation was implemented in this analysis. Several techniques are used to measure the quality of terrain elevation models (Wheaton et al., 2010), which include: the accuracy given by the equipment manufacturer, the repetition of observations using control points, bootstrapping methods, fuzzy terrain models, repeated surveys of areas without any change and other geostatistical techniques.

In this investigation we used the Minimum Level of Detection (LOD_{Min}). δU_{DoD} is a constant that represents the combined error arising from the addition or subtraction of two DEMs, $Z_{(1)}$ and $Z_{(2)}$. This constant can be estimated from the root sum square of errors:

$$\delta U_{DoD} = \sqrt{\delta z_{(1)}^2 + \delta z_{(2)}^2} \quad (1)$$

where $\delta z_{(1)}$ and $\delta z_{(2)}$ represent the errors related with $Z_{(1)}$ and $Z_{(2)}$, respectively. For example, if $\delta z_{(1)}$ and $\delta z_{(2)}$ is 0.1 m in each DEM, then δU_{DoD} has a value of 0.14 m. To use the LOD_{Min}, the value of δU_{DoD} is applied in the DoD analysis as a constant threshold. This method is conservative because only the geomorphic change that is greater than the LOD_{Min} is considered adequate. In fact, analyses by Brasington et al. (2003) and Wheaton et al. (2010) indicate that areal and volumetric estimates of morphological change are highly sensitive to this threshold, i.e. information about the actual geomorphic change is very likely to be lost below this threshold (Williams, 2012). An ArcGIS extension for analyzing geomorphic change (Geomorphic Change Detection, http://gcd.joewheaton.org/home) was installed as a toolbox within the ArcMap graphical interface. This tool includes a series of procedures such as: the preparation phase of the DEMs and the application of change detection using various methods. The output formats are ESRI GRID files, graphics and txt files.

3.2 Bedform geometry

A GPSmap 62S GARMIN was used to analyze river bed elevation changes. Changes as small as 0.001 m can be measured with this equipment to evaluate the bedform geometry. The depth sounder followed a path with the streamwise direction.

Two longitudinal transects were measured and repeated again 24 h later. The length and lateral separation of the two sounding lines (orange lines) are shown in Fig. 1a. The acquisition of the bed profiles was validated directly in the field using the equipment graphical interface and later the bathymetric points were exported and mapped on a defined fixed

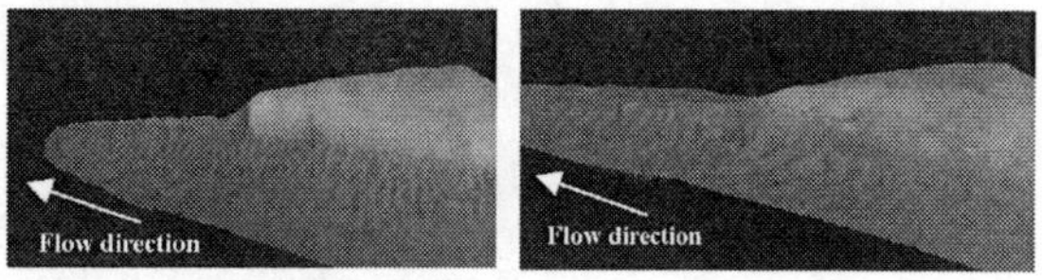

Figure 4. Small bed features (sand dunes) were mapped on the bottom of the river using ADCP repeated transects acquired at different stages of the hydrological regime.

scale. Sand waves and bars can be seen in each cross section. To calculate the height of each dune the difference in elevation between the lowest and highest points of the dune was determined. The slope of the dune was another parameter evaluated, calculated as the ratio of the height and the length of the dune.

Comparative graphs between repeated transects every 24 h were performed in order to understand the dynamics of the sand dunes in the study reach, as well as to have an approximate idea of the rate of displacement of these bed forms at different flow conditions.

4 Results

4.1 Morphological changes recorded by DEM of Difference (DoD) over a mid-channel bar

The low errors associated with the DEMs, derived from the combination of RTK-DGPS measurements and repeated ADCP transects around the island (mid-channel bar), suggest that the methodology is capable of delivering high-accuracy topographic and bathymetric models.

Results of a single ADCP transect are insufficient to build adequate bottom bathymetry. However, the redundancy of multiple ADCP transects on the same path allows the acquisition of a large number of bed elevation points. This allows a sufficiently detailed and precise bathymetric surface to be constructed. Each ADCP transect was separated from the previous path by an appropriate distance following the path of the graphical interface of the GPS. This generated a distribution of thousands of bed elevation points around the mid-channel bar. Subsequently, the points were interpolated and converted into a raster model, thus obtaining the bathymetric surface of the sector with a spatial resolution of 1 m. The reliability of this methodology using repeated transects was confirmed by the observation of small bed forms in the 3-D model of the bathymetric surface. This procedure was repeated at different flow conditions: in November 2016 (during falling waters), and at the start of rising waters in April 2017 (Fig. 4).

As shown in Fig. 5, during the time elapsed between these two surveys, it was possible to analyze mean flow velocity vectors in both periods. A marked decrease of the river flow

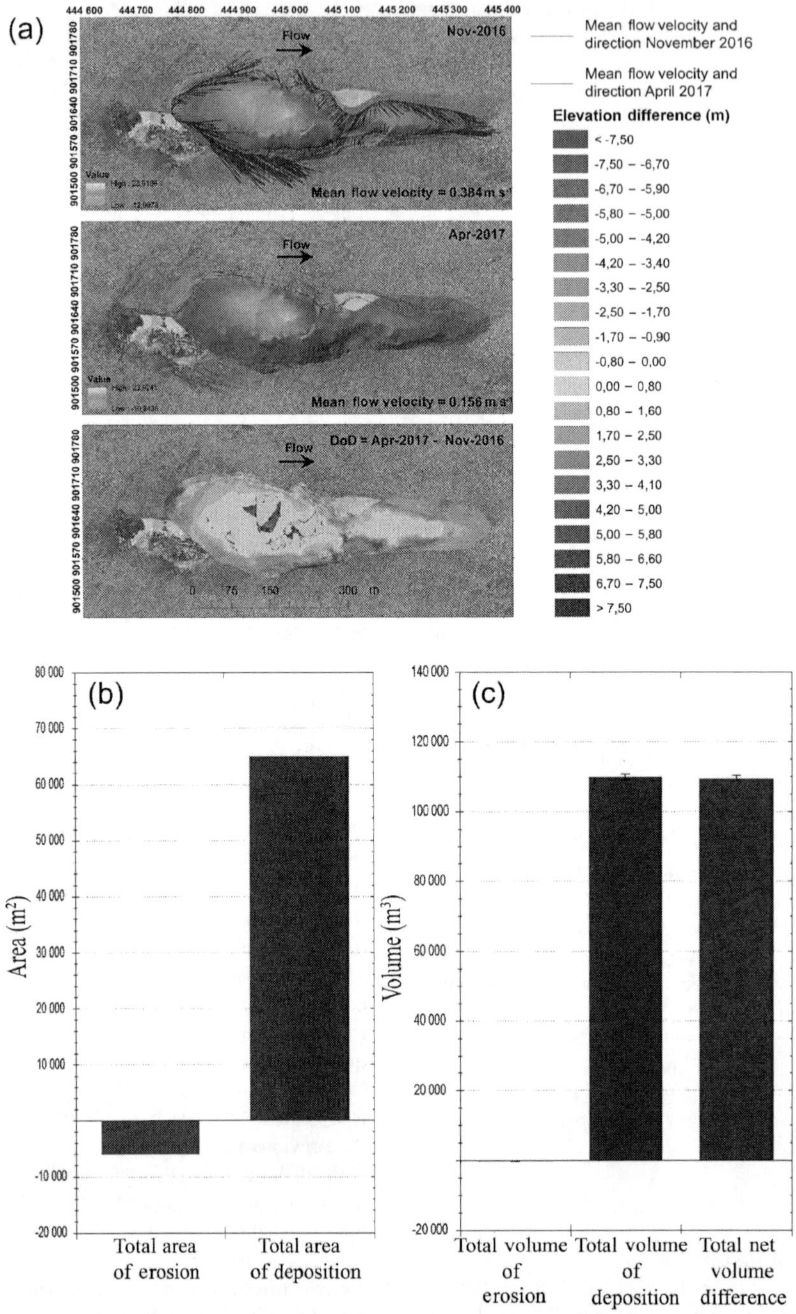

Figure 5. **(a)** DEMs on the two study dates and DoD over the mid-channel bar using $LOD_{min} = 0.01$. In the DoD the red color represents erosion and the blue color deposition. **(b, c)** show the morphological changes in area (m^2) and volume (m^3).

velocity between the two survey dates occurs, from 0.384 to $0.156\,m\,s^{-1}$. The morphological change resulting from the integration of these events can be revealed by subtracting the two (after and before) surface models to generate a DEM of difference. During this study period deposition of sediment occurs on both flanks of the mid-channel bar. This is due to falling water levels that result in deposition of much of the suspended sediment load over the island, promoting vertical growth by sediment accumulation, especially in the sand dune located downstream of the island.

The results from the DoD analysis are shown in Table 2. The total area of sediment along the mid-channel bar was calculated at about 65 037 m^2, while the total area of erosion was 6066 m^2 during this study period (Fig. 5).

The total volume of sediment was $109\,792 \pm 911\,m^3$ representing 99.67 % of the total net volume of difference, while the total volume of erosion was $368 \pm 85\,m^3$, which represents 0.33 % during the study period. An error assessment was implemented using a $LOD_{Min} = 0.01$ calculated by Eq. (1). This value was based on the errors from the manufacturer reported instrument precision used to generate $\delta z_{(1)}$ and $\delta z_{(2)}$. To implement the LOD_{Min}, the value of δU_{DoD} was applied as a constant threshold across the DoD.

To illustrate the evolution of the mid-channel bar during the period of falling waters a 3-D model was generated showing the topography of the sand bar and surrounding bathymetry at different stages of the river (Fig. 6). The formation of this mid-channel bar is likely related to the complex interaction between the change in discharge and water level in the river, the subsequent adjustment of flow velocity/direction and the sediment transport. Figure 6b shows the distribution of the flow velocity vectors surrounding the mid-channel bar. Here, with the advance of the high flow velocity area downstream, especially during the high water period, material is eroded from the tail of the bar. Eventually, this material is deposited the foot of the front slope of the sand bar, where it accumulates.

In Fig. 6c, the discharge decreases and most of the sediment load in suspension is deposited promoting vertical growth. This change in sedimentation volume of the middle channel bar is directly related to the bedload in motion. Analysis of these changes at different river stages may allow us to understand better the dynamics of sediment transport in the bed.

A multimedia file of this simulation is available for downloading from Yepez (2017).

One of the most important aspects of this analysis is that island and bank heights, obtained by repeated ADCP transects in combination with RTK-DGPS measurements, allows the estimation of volume rates of erosion and deposition in an easy to implement manner. Considering that many of the banks are submerged in many sections of the river as well as in the islands, these techniques are suitable for characterizing the submerged portion of the banks and the riverbed.

Although this paper focuses on the estimation of accretion and erosion volumes in this mid-channel bar over just one year, it is necessary to evaluate the same stages of the river in different years, in order to understand changes in the sediment balance throughout the hydrological regime. This is the focus of ongoing study.

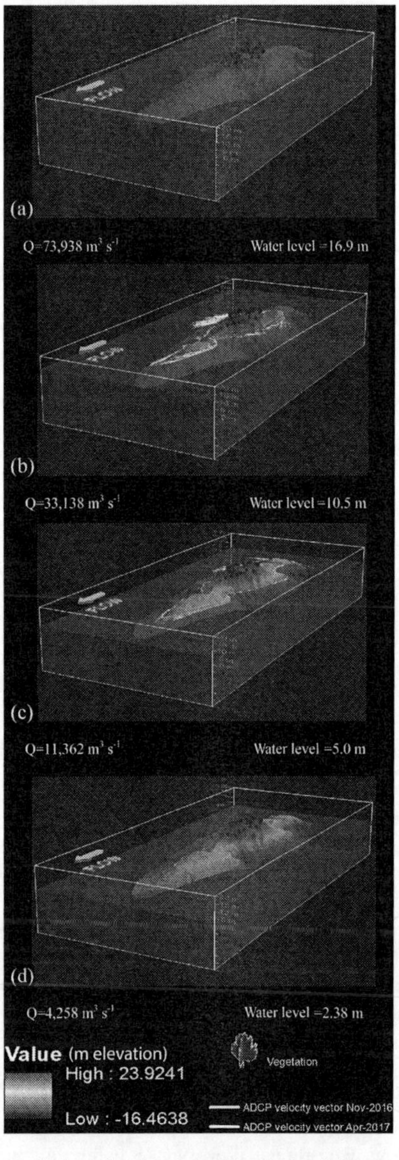

Figure 6. Evolution of a mid-channel bar during falling waters. The yellow **(b)** and white **(c)** lines that surround the island represent the flow velocity vectors for the periods of high and low waters, respectively.

4.2 Kinematics of sand dunes at Ciudad Bolivar

The results obtained from the comparison between the May 2016 and November 2016 surveys in this study reach are shown in Fig. 7. In this analysis it is possible to appreciate significant variations in the average rates of migration of the dunes in relation to the discharge. During the rising waters period there was a greater difference between the bathymetry

Table 2. Summary of DoD analysis.

Attribute	Raw	Thresholded DoD Estimate	
Areal			
Total Area of Erosion (m^2)	7753	6066	
Total Area of Deposition (m^2)	67 123	65 037	
Total Area of Detectable Change (m^2)	NA	71 103	
Total Area of Interest (m^2)	74 876	NA	
Percent of Area of Interest with Detectable Change	NA	94.96 %	
Volumetric		± Error Volume	% Error
Total Volume of Erosion (m^3)	379.77	368 ± 85	23.07 %
Total Volume of Deposition (m^3)	109 807	109 792 ± 911	0.83 %
Total Volume of Difference (m^3)	110 187	110 159 ± 995	0.90 %
Total Net Volume Difference (m^3)	109 427	109 423 ± 914	0.84 %

of the two profiles taken 24 h apart compared to during the falling water period.

The highest migration rates ($0.07\,\mathrm{cm\,s^{-1}}$) occur during periods of rising waters compared to the stages of falling waters when migration rates decrease. It is observed that if the dunes are presented with a high frequency and with small sizes during the period of rising waters, the rates will be high. In contrast, in the falling waters stage, the transport of sediment at the same rate in larger dunes implies slower migration rates.

5 Conclusions and recommendations

The volume of deposition/erosion over the mid-channel bar, and their interrelationship with discharge in the Orinoco River was estimated through processing from DEM of Difference (DoD) maps on time series of bathymetric and topography data. Likewise, the analysis of longitudinal bathymetric transects 24 h apart, allowed investigation of the evolution and the kinematics of the sand dunes in this study reach. The results of this study highlighted the effect of seasonal floods on bank accretion/erosion in this fluvial island.

Furthermore, several bathymetric transects perpendicular to the flow were repeated every 24 h, which allowed understanding of the migration rates of sand waves and dunes in the study section. These comparisons made in relatively short periods of time facilitated the analysis of the kinematics of the sand dunes, in order to get a clearer idea about the evolution of these bed forms at different flow conditions.

This paper presents a detailed methodology for the collection and processing of data associated with numerical terrain models, allowing measurements of morphological changes of a fluvial island, which is representative of the lower Orinoco, one of the most important rivers in South America and worldwide.

From the results of DoD analysis to quantify the deposition/erosion volume of this mid-channel bar, it was possible to determine that during the time between the two surveys, sedimentation played a more important role than erosion. Also, evaluating the vectors of mean flow velocity in both periods, a marked decrease from 0.384 to $0.156\,\mathrm{m\,s^{-1}}$ was observed. When the discharge began to decrease, part of the sediment load in suspension was subjected to deposition causing a vertical growth of the sediments on the bar, especially in the most downstream part of the island, where a sand dune developed in a needle shape.

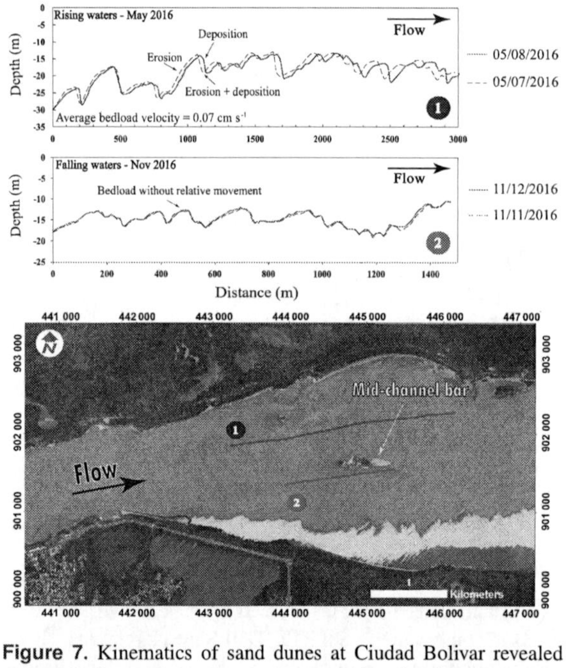

Figure 7. Kinematics of sand dunes at Ciudad Bolivar revealed through bathymetric profiles (upper figures) and a plan view of the profile transects (lower figure). (1) corresponds to rising waters, while (2) corresponds to falling waters.

One of the aspects that will have to be addressed in future work will be the sedimentary balance. For this, it is necessary to evaluate the same periods of time in different years to understand the morphological changes during the variations in the hydrological regime. The next field campaigns will focus on these objectives. The quality of the DEMs obtained with the implementation of this combined methodology of ADCP transects and RTK-DGPS measurements allowed high-resolution bathymetric models to be created, where it is possible to observe the sand-wave patterns.

Enhancing knowledge of the role of morphological changes (volume) with this new methodology is an opportunity to gain better understanding of river bed sediment transport. This type of study will support dredging projects in the Orinoco River to maintain navigability, which will contribute to the management of this important river basin.

Data availability. A multimedia file of this simulation is available for downloading from Yepez (2017).

Competing interests. The authors declare that they have no conflict of interest.

Special issue statement. This article is part of the special issue "Water quality and sediment transport issues in surface water". It is a result of the IAHS Scientific Assembly 2017, Port Elizabeth, South Africa, 10–14 July 2017.

Acknowledgements. The field campaigns on the Orinoco River were supported by the SO-HYBAM project and ECOS-Nord/Fonacit (V14U01), as well as by the Bolivarian National Armada of Venezuela and COEA-IVIC laboratory. The first author was financially supported by IRD-ARTS Grant 2017-2018 and Venezuela's Fundayacucho Grant No. E-223-14-2014-2. The paper greatly benefited from the thorough reviews of anonymous referees.

Edited by: Kate Heal
Reviewed by: two anonymous referees

References

Acuña, G.: VGM 17, el nuevo modelo geoidal LGFS-LUZ de ultra-alta-resolución 30 × 30 m para Venezuela y regiones vecinas, Notas de Geodesia Geométrica, Laboratorio de Geodesia Física y Satelital, Dpto. de Geodesia Superior, Escuela de Ingeniería Geodésica, Facultad de Ingeniería, La Universidad del Zulia, February 2017, in preparation, available at: http://ggenluz.blogspot.fr/2017/01/comienza-la-produccion-de-vgm17-el.html (last access: 5 February 2018), 2017.

Brasington, J., Langham, J., and Rumsby, B.: Methodological sensitivity of morphometric estimates of coarse fluvial sediment transport, Geomorphology, 53, 299–316, https://doi.org/10.1016/S0169-555X(02)00320-3, 2003.

Carrivick, J. L., Geilhausen, M., Warburton, J., Dickson, N. E., Carver, S. J., Evans, A. J., and Brown, L. E.: Contemporary geomorphological activity throughout the proglacial area of an alpine catchment, Geomorphology, 188, 83–95, https://doi.org/10.1016/j.geomorph.2012.03.029, 2013.

Depetris, P. J. and Paolini, J.: Biogeochemical aspects of South American rivers: The Paraná and the Orinoco, in: Biogeochemistry of Major World Rivers, edited by: Degens, E. T., Kempe, S., Richey, J., J. Wiley & Sons, Chichester, 105–125, 1991.

Dinehart, R. and Burau, J.: Repeated surveys by acoustic Doppler current profiler for flow and sediment dynamics in a tidal river, J. Hydrol., 314, 1–21, https://doi.org/10.1016/j.jhydrol.2005.03.019, 2005.

Fuller, I., Passmore, D., Heritage, G., Large, A., Milan, D., and Brewer, P.: Annual sediment budgets in an unstable gravel-bed river: the River Coquet, northern England, Geological Society, London, Special Publications, 191, 115–131, https://doi.org/10.1144/GSL.SP.2002.191.01.08, 2002.

Laraque, A., Castellanos, B., Steiger, J., Lopez, J. L., Pandi, A., Rodriguez, M., Rosales, J., Adèle, G., Perez, J., and Lagane, C.: A comparison of the suspended and dissolved matter dynamics of two large inter-tropical rivers draining into the Atlantic Ocean: the Congo and the Orinoco, Hydrol. Process., 27, 2153–2170, https://doi.org/10.1002/hyp.9776, 2013.

Lawler, D., Grove, J., Couperthwaite, J., and Leeks, G.: Downstream change in river bank erosion rates in the Swale-Ouse system, northern England, Hydrol. Process., 13, 977–992, https://doi.org/10.1002/(SICI)1099-1085(199905)13:7<977::AID-HYP785>3.0.CO;2-5, 1999.

Lewis Jr., W. M. and Saunders III, J. F.: Cross-sectional variation in the chemistry and suspended sediment load of the Orinoco River at Ciudad Bolívar, Acta Cientifica Venezolana, 35, 382–385, 1984.

Lewis Jr., W. M. and Saunders III, J. F.: Concentration and transport of dissolved and suspended substances in the Orinoco River, Biogeochemistry, 7 , 203–240, 1989.

Leyland, J., Hackney, C. R., Darby, S. E., Parsons, D. R., Best, J. L., Nicholas, A. P., Aalto, R., and Lague, D.: Extreme flood-driven fluvial bank erosion and sediment loads: direct process measurements using integrated Mobile Laser Scanning (MLS) and hydro-acoustic techniques, Earth Surf. Proc. Land., 42, 334–346, https://doi.org/10.1002/esp.4078, 2016.

López, J. and Perez-Hernandez, D.: Some Morphological Aspects of the Orinoco River, in: IAHR Symposium on River, Coastal and Estuarine Morphodynamics, Genova, Italy, 6–10 September 1999.

Mat Salleh, M. and Ariffin, J.: Flow and Sediment Matrix in Mid-Channel Bar Formation, Int. J. Sci. Eng. Res., 4, 1757–1764, 2013.

Paolini, J., Hevia, R., and Herrera, R.: Transport of carbon and minerals in the Orinoco and Caroni rivers during the years 1983–1984, Mitteilungen aus dem Geologisch-Paläontologischen Institut der Universität Hamburg, 64, 325–338, 1987.

Parsons, D., Jackson, P., Czuba, J., Engel, F., Rhoads, B., Oberg, K., Best, J., Mueller, D., Johnson, K., and Riley, J.: Velocity Mapping Toolbox (VMT): a processing and visualization suite for moving-vessel ADCP measurements, Earth Surf. Proc. Land., 38, 1244–1260, https://doi.org/10.1002/esp.3367, 2013.

Silva León, G.: La cuenca del río Orinoco: visión hidrográfica y balance hídrico (The Orinoco River basin: hydrographic view and its hydrological balance), Revista Geografica Venezolana, 46, 75–108, 2005.

Warne, A. G., Meade, R. H., White, W. A., Guevara, E. H., Gibeaut, J., Smyth, R. C., Aslan, A., and Tremblay, T.: Regional controls on geomorphology, hydrology, and ecosystem integrity in the Orinoco Delta, Venezuela, Geomorphology, 44, 273–307, https://doi.org/10.1016/S0169-555X(01)00179-9, 2002.

Wheaton, J. M., Brasington, J., Darby, S. E., and Sear, D. A.: Accounting for uncertainty in DEMs from repeat topographic surveys: improved sediment budgets, Earth Surf. Proc. Landf., 35, 136–156, https://doi.org/10.1002/esp.1886, 2010.

Williams, R.: DEMs of difference, Geomorphological Techniques, vol. 2, no 3.2, 2012.

Williams, R., Brasington, J., Vericat, D., Hicks, M., Labrosse, F., and Neal, M.: Chapter twenty-monitoring braided River change using terrestrial laser scanning and optical bathymetric mapping, Dev. Earth Surf. Proc., 15, 507–532, https://doi.org/10.1016/B978-0-444-53446-0.00020-3, 2011.

Yepez, S., Laraque, A., Martinez, J. M., De Sa, J., Carrera, J. M., Gallay, M., and Lopez, J. L.: Retrieval of suspended sediment concentrations using LANDSAT-8 OLI data in the Orinoco River (Venezuela), Journée Thématique du Programme National de Télédétection Spatiale: Apport des missions Sentinel – Copernicus à l'observation de la Terre, CNES, Paris, 24 March, https://doi.org/10.13140/RG.2.2.36296.32005, 2016.

Yepez, S.: Evolution of a mid-channel bar during falling waters (Orinoco River – Venezuela), https://doi.org/10.13140/RG.2.2.14443.64802, 2017.

Proc. IAHS, 377, 51–56, 2018
https://doi.org/10.5194/piahs-377-51-2018
© Author(s) 2018. This work is distributed under
the Creative Commons Attribution 4.0 License.

Morphodynamic simulation of sediment deposition patterns on a recently stripped bedrock anastomosed channel

David Milan[1], **George Heritage**[2], **Neil Entwistle**[3], **and Stephen Tooth**[4]

[1]School of Environmental Science, University of Hull, Cottingham Road, Hull, HU6 7RX, UK
[2]AECOM, Exchange Court, 1 Dale Street, Liverpool, L2 2ET, UK
[3]School of Environment and Life Sciences, University of Salford, Salford, M5 4WT, UK
[4]Department of Geography and Earth Sciences, Aberystwyth University, Aberystwyth, SY23 3DB, UK

Correspondence: David Milan (d.milan@hull.ac.uk)

Received: 9 June 2017 – Accepted: 16 August 2017 – Published: 16 April 2018

Abstract. Some mixed bedrock-alluvial dryland rivers are known to undergo cycles of alluvial building during low flow periods, punctuated by stripping events during rare high magnitude flows. We focus on the Olifants River, Kruger National Park, South Africa, and present 2-D morphodynamic simulations of hydraulics and sediment deposition patterns over an exposed bedrock anastomosed pavement. We examine the assumptions underlying a previous conceptual model, namely that sedimentation occurs preferentially on bedrock highs. Our modelling results and local field observations in fact show that sediment thicknesses are greater over bedrock lows, suggesting these are the key loci for deposition, barform initiation and island building. During peak flows, velocities in the topographic lows tend to be lower than in intermediate topographic areas. It is likely that intermediate topographic areas supply sediment to the topographic lows at this flow stage, which is then deposited in the lows on the falling limb of the hydrograph as velocities reduce. Subsequent vegetation establishment on deposits in the topographic lows is likely to play a key role in additional sedimentation and vegetation succession, both through increasing the cohesive strength of alluvial units and by capturing new sediments and propagules.

1 Introduction

Many of the large rivers draining southern Africa are characterised by bedrock "macrochannels" incised 10–20 m into ancient planation surfaces but with variable amounts of unconsolidated sediment infill. Variations in lithology, structure, flow regime, sediment supply, and vegetation assemblages have created morphologically diverse river systems. Many rivers have been characterised by extended periods of alluviation on an historic timescale (Tooth, 2016), leading to the development of increasingly alluvial channel types with the underlying bedrock commonly masked. Alluviation is periodically interrupted by extreme flood stripping of accumulated sediment (Heritage et al., 2001), events which have also been highlighted as an important mechanism behind fluvial change and system resetting in Australian, Indian and North American river systems (e.g. Baker, 1977; Nanson, 1986; Kale et al., 1996).

Anastomosed channels represent one of the key channel types in many mixed bedrock-alluvial rivers in the Kruger National Park (KNP), South Africa (Heritage et al., 2004). Response in these channel types is largely controlled by different magnitude flood events; alluvium and vegetation is stripped from the bedrock template during large or extreme events, and intervening building phases with lower magnitude floods allow sediment accumulation and vegetation establishment through plant succession (van Niekerk et al., 1999; Rountree et al., 2000). In van Niekerk et al.'s (1999) conceptual model, during building phases, flood deposition occurs preferentially on topographic highs on the stripped bedrock template, initiating sedimentation and vegetation establishment, while topographic lows remain relatively sediment free. To date, there has been little empirical evidence available to enable examination of the assumptions underlying this conceptual model. Stripping events tend to be rare, and obtaining field data on sedimentation can be difficult due to the presence of dangerous animals, dense vegetation and the large scale of the channels. Using morphodynamic modelling and field observations, this paper investigates the processes driving alluviation and bar/island development over a bedrock template, examining in particular the influence of topography on hydraulics and depositional loci.

2 Study site

The Olifants River is located in the southern and central part of the KNP in the Mpumalanga and Limpopo provinces of north-eastern South Africa (Fig. 1). The 54 570 km^2 Olifants catchment covers part of the Highveld Plateau (~ 2000–1500 m a.s.l.), the Drakensberg Escarpment, and the Lowveld (~ 400–250 m a.s.l.). Rainfall is greater in the highlands (~ 2000 mm yr^{-1}) and declines rapidly eastwards towards the South Africa–Mozambique border (~ 450 mm yr^{-1}). Although water abstractions have altered the low flows (typically < 50 m^3 s^{-1}), flood flows are unaffected, and the river remains unimpacted by engineering structures or other human activities over considerable lengths within the KNP. Thus, the river provides excellent, near-pristine sites for investigating dryland, bedrock-influenced channel dynamics.

The Olifants River in the KNP is characterised by a bedrock macrochannel up to ~ 500 m wide that is incised 10–20 m into the Lowveld. Outside of the macrochannel, floods have a very infrequent and limited influence. The river is characterised by a high degree of bedrock influence, but the bedrock may be buried by alluvial sediments of varying thickness. The diverse underlying geology (principally basalt, with some tonalitic gneiss) results in frequent, abrupt changes in macrochannel slope and associated sediment deposition patterns, creating diverse channel types that include alluvial single-thread and braided channels, and bedrock anastomosed and pool-rapid channels (van Niekerk et al., 1995; Milan et al., 2017).

A large cyclone-driven flood event occurred in January 2012, resulting in an estimated peak discharge of between 14 407 to 16 772 m^{3}s^{-1} (Heritage et al., 2017). This flood led to extensive stripping of alluvium from the channels to reveal the underlying bedrock template. Our study focused on a 4 km length of stripped bedrock anastomosed channel (Fig. 1c) as this provided an opportunity to examine the processes responsible for the initiation of deposition over the stripped bedrock following the flood.

3 Methods

Aerial LiDAR and orthophotography was obtained on 30 May 2012 for a 50 km reach of the lower Olifants in the KNP. Southern Mapping Geospatial surveyed the river using an Opetch Orion 206 LiDAR, and a Rollei AIC with a 60 mega-pixel P65+ Phase One digital CCD, flown from a Cessna 206 at 1100 m altitude. To allow 2-D morphodynamic modelling, the LiDAR was degraded to 5 m resolution, which is typically smaller than the width of the anabranches through the bedrock. A field visit was also conducted in May 2012 to gather evidence of processes influencing fine sediment deposition and reworking along the river.

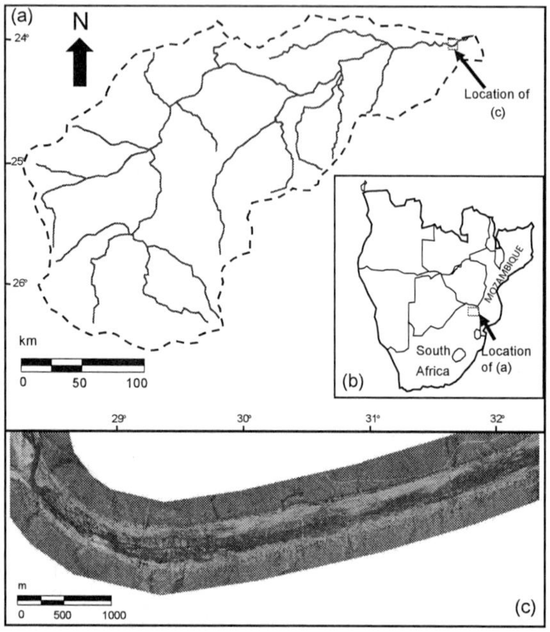

Figure 1. Olifants River catchment; **(a)** catchment map, **(b)** location in South Africa, **(c)** aerial photograph (May 2012) of reach in the Kruger National Park used for simulation.

2-D flow and sediment transport modelling was undertaken to investigate hydraulic and sediment transport processes more closely for the study reach using the 2012 LiDAR data. We used the CAESAR Lisflood morphodynamic code (Coulthard et al., 2013), with a 14 day flow time series (Fig. 2a) that included two hydrograph peaks (4500 and 2000 m^3 s^{-1}), sufficient to fully submerge bedrock highs in the channel. The model was run on the bare earth 5 m DEM with a Mannings "n" of 0.04, and was set to recirculate any eroded sediment. For alluvial channels, bedrock is typically at an elevation lower than the surface elevation, but here the bedrock was set to the level of the surface to produce a resistant macrochannel template. A slug of sand-size sediment ($D_{50} = 1.4$ mm) was introduced from upstream with a feed rate of between 2×10^5 and 90×10^5 m^3 h^{-1} to reflect the fact that flows along the river are generally transport- rather than supply-limited. The aim was to investigate stage-dependent hydraulics and patterns of erosion and deposition relative to the bedrock template. Emphasis was placed on mapping

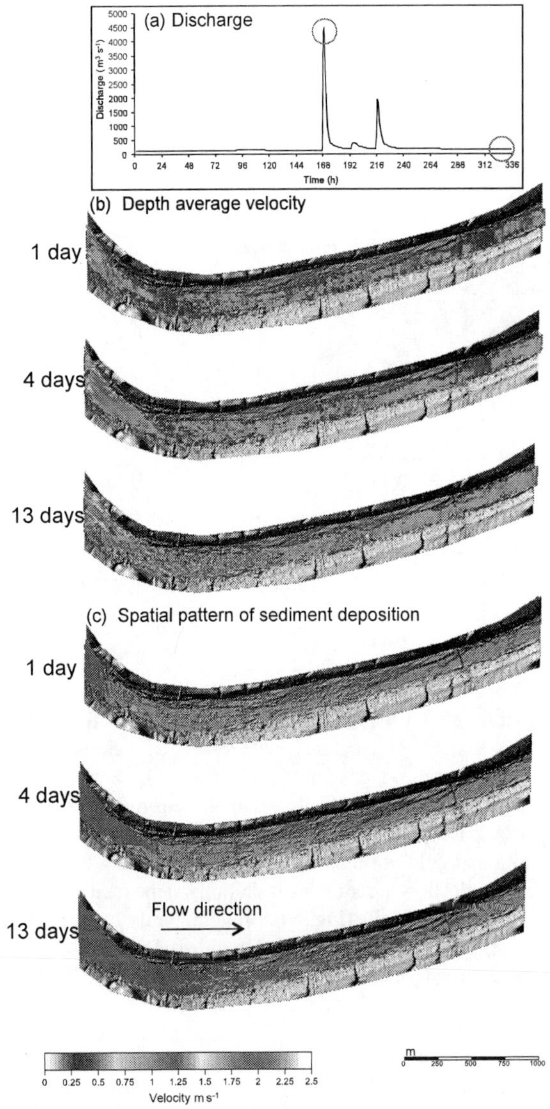

Figure 2. CAESAR-Lisflood simulations: **(a)** Discharge time-series with 4500 and 200 $m^3 s^{-1}$ flows highlighted; **(b)** depth-averaged velocity patterns, **(c)** spatial patterns of deposition as a sediment slug (orange shading) is dispersed through the model domain.

depositional loci and identifying their topographic and hydraulic controls. For each 5 m pixel on the DEM, comparisons were made between the detrended bedrock template elevations and: (1) sediment thicknesses deposited at the end of the model run; (2) velocities for a high (4500 $m^3 s^{-1}$) and low (200 $m^3 s^{-1}$) discharge.

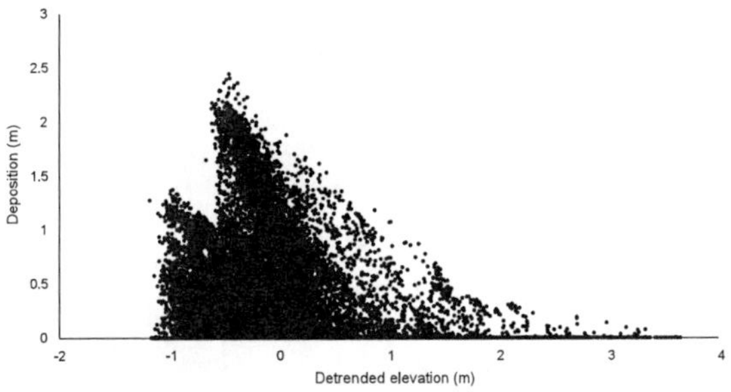

Figure 3. Sediment thicknesses for each 5 m pixel of study reach versus detrended bed elevation.

4 Results

4.1 Spatial patterns of deposition and channel development

Three snapshots taken over the 14 day duration of the simulation are shown in Fig. 2b–c. Early in the simulation, flow is running predominantly over the bedrock template, and for the lower discharges is controlled by the anastomosed channel network etched into the bedrock. Gradually, the sediment that is fed into the model is distributed downstream, most markedly following the peak discharge of 4500 $m^3 s^{-1}$. After 13 days, the upstream most 800 m of the reach is fully alluvial, and a braided channel network is evident, largely independent of the anabranches in the bedrock beneath (Fig. 2c; 13 days). Although it is likely that the model run time was insufficient to fully disperse the slug supplied to the reach, field observations show that this channel type is becoming more common on the river and these results suggest that superficial braiding will persist in the short to medium term. Farther downstream, patches of sediment clearly accumulate as lobate barforms (Fig. 2c; 13 days); field observations show that such features are indeed present along the river and are becoming colonised by *Phragmites* spp.

4.2 Simulated sediment thicknesses

Sediment thicknesses for each 5 m pixel of a detrended raster DEM covering the barforms and exposed bedrock areas were extracted. To identify the association between sediment thickness and topography, the bedrock template DEM was detrended to remove any downstream slope trend. Detrended elevations for each 5 m pixel were then retrieved for the same loci as the sediment thicknesses.

The resultant plot (Fig. 3) clearly shows greater sediment accumulation in areas of negative bed topography i.e. topographic lows. The sediment thickness data therefore appear to suggest that these barforms are initiated by the infilling of bedrock topographic lows.

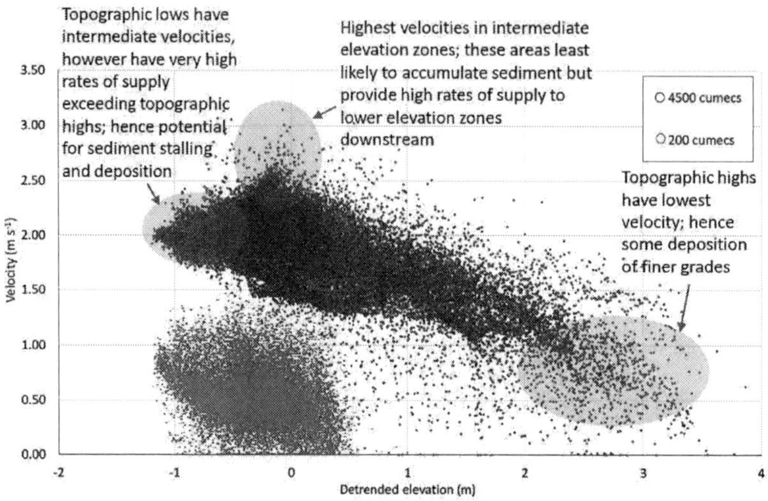

Figure 4. Velocity data taken from the same 5 m pixels as the sediment thickness data (Fig. 3): for low discharge (red: $200\,m^3\,s^{-1}$), and high discharge (black: $4500\,m^3\,s^{-1}$).

4.3 Topographic controls on velocity patterns

To explore the controls of topography upon flow hydraulics, depth averaged velocities for the same 5 m pixels used to derive the sediment thickness data were extracted for a competent but relatively low flow discharge of $200\,m^3\,s^{-1}$ and a high flow of $4500\,m^3\,s^{-1}$. These discharges are representative of the typical range of floods that occur on decadal timescales, during which net sedimentation (as opposed to net erosion) occurs. The relationship between topography and simulated depth-averaged velocities was investigated for these different flow conditions (Fig. 4). During the high flow ($4500\,m^3\,s^{-1}$), topographic highs are only just submerged and have the lowest velocity ($<1\,m\,s^{-1}$), which is likely to promote deposition of fine sediment. The highest velocities (2–$3\,m\,s^{-1}$) are found for intermediate elevation zones; these areas are least likely to accumulate sediment but probably facilitate sediment supply to lower elevation zones downstream. Topographic lows also have moderate velocities ($\sim 2\,m\,s^{-1}$), which are capable of entraining sand and even gravel clasts if available, so are not likely to accumulate fine sediment during peak flows. For the low flow ($200\,m^3\,s^{-1}$), topographic highs are exposed, with no flow occurring over these areas (Fig. 4). Velocity reductions are dramatic as discharge falls, most notably in lower topographic zones (Fig. 4), which promotes rapid deposition of fines and infilling of bedrock hollows. The intermediate elevation zones are submerged at low flow and have a similar velocity range to lower elevation areas. Most of the flow is concentrated into the topographic lows, however, where sediment is likely to be routed and reworked, but with deposition occurring in lower energy areas.

5 Discussion

Heritage et al. (2014) have presented OSL dates for the KNP rivers that suggest extensive flood-related stripping events tend to occur on centennial timescales. In between these extreme flood events, sediment builds up over the stripped bedrock template during lower magnitude flows. Our modelling and field observations indicate that during these lower magnitude flows, sediment is initially deposited rapidly as flood discharge falls (Fig. 5a), predominantly in topographic lows (Fig. 5b) where some sediment may be sheltered from erosion in subsequent floods. These findings represent a refinement to van Niekerk et al.'s (1999) more generalised conceptual model that assumed near-uniform, ubiquitous sedimentation over bedrock highs, and are supported by the flume investigations of Hodge and Hoey (2016). An additional field observation is that fine sediment (e.g. silt) that is deposited away from the low flow channel network may be eroded by rainfall, and reworked into topographic lows (Fig. 5c). This augments the flood deposits and helps promote colonisation by early successional vegetation such as *Phragmites mauritianus*.

Once vegetation establishes, then the sediment unit is strengthened owing to the development of root structures. Vegetation also helps to trap sediment and propagules supplied from upstream, encouraging further alluviation and plant growth. For instance, growth of *Breonadia salacina* in bedrock discontinuities and later successional *Combretum* spp. and *Syzgium guineense* enables the capture of organic material during floods and provides potential for further vegetation development from these nuclei. Sediment thicknesses are likely to develop further in what were originally bedrock

Figure 5. Sediment deposition patterns on the Olifants River in the Kruger National Park: **(a)** post-flood peak silt drapes (flow direction from left to right); **(b)** sand accumulation in topographic lows (flow direction from top to bottom); **(c)** sand deposition in topographic lows and areas in the lee of bedrock outcrop (flow direction from right to left). Silt deposition is also evident in the lee of the outcrop (see foreground), and in the stoss side hollow, possibly owing to reworking from the topographic high by rainfall.

topographic lows, eventually leading to bedrock core bars and islands, with the alluviated high loci situated over the bedrock topographic low loci. In this way, many of the bedrock low points have the thickest and wettest substrate for plant development, and are less susceptible to complete stripping during large floods. Field observations based on "before" and "after" comparisons of extreme floods (Milan et al., 2017) show that many topographic highs have a tendency to be stripped during flood events capable of submerging these areas. We intend to further test these hypotheses through running more simulations on other sites with different channel types, and also using repeat LiDAR data for sites that were stripped in the 2012 event.

6 Conclusions

1. Barforms are initiated by sediment deposition that occurs over preferentially in bedrock topographic lows. These lows display mid-range velocities during the peak of high flow events, but tend to be preferential routing zones for sediment, with deposition occurring in low energy areas on the falling limb of floods;
2. Although some initial deposition of fine sediment can occur on topographic highs following floods, some may be reworked into topographic lows by rainfall;
3. Vegetation development is likely to be fundamental to subsequent barform and island growth as it increases the cohesive strength of sediment and captures additional sediment and propagules, but this needs to be tested by additional modelling work.

Data availability. The authors do eventually intend to make the LiDAR data available through the NERC data repository.

Competing interests. The authors declare that they have no conflict of interest.

Special issue statement. This article is part of the special issue "Water quality and sediment transport issues in surface water". It is a result of the IAHS Scientific Assembly 2017, Port Elizabeth, South Africa, 10–14 July 2017.

Acknowledgements. This project was funded through the Natural Environment Research Council Urgency Grant NE/K001132/1. We would like to thank SANParks for supporting this research.

Edited by: Gil Mahe
Reviewed by: Jens Turowski and Rebecca Hodge

References

Baker, V. R.: Stream-channel response to floods, with examples from central Texas, Geol. Soc. Am. Bull., 88, 1057–1071, 1977.

Coulthard, T. J., Neal, J. C., Bates, P. D., Ramirez, J., Almeida, G. A., and Hancock, G. R.: Integrating the LISFLOODFP 2D hydrodynamic model with the CAESAR model: implications for modelling landscape evolution, Earth Surf. Proc. Landf. 38, 1897–1906, 2013.

Heritage, G., Tooth, S., Entwistle, N., and Milan, D.: Long-term flood controls on semi-arid river form: evidence from the Sabie and Olifants rivers, eastern South Africa, Proc. IAHS, 367, 141–146, https://doi.org/10.5194/piahs-367-141-2015, 2014.

Heritage, G. L., Moon, B. P., Jewitt, G. P., Large, A. R. G., and Rountree, M.: The February 2000 floods on the Sabie River, South Africa: an examination of their magnitude and frequency, Koedoe, 44, 37–44, 2001.

Heritage, G. L., Large, A. R. G., Moon, B. P., and Jewitt, G.: Channel hydraulics and geomorphic effects of an extreme flood event on the Sabie River, South Africa, Catena, 58, 151–181, 2004.

Heritage, G. L., Entwistle, N., Milan, D. J., and Tooth, S.: Estimating flood magnitude using a 2D modelling approach: examples from the Sabie and Olifants rivers, eastern South Africa, Catena, in review, 2017.

Hodge, R. A. and Hoey, T. B.: A Froude-scaled model of a bedrock alluvial channel reach: 2. Sediment cover, J. Geophys. Res.-Earth Surf., 121, 1597–1618, 2016.

Kale, V. S., Baker, V. R., and Mishra, S.: Multi-channel patterns of bedrock rivers: an example from the central Narmada basin, India, Catena, 26, 85–98, 1996.

Milan, D. J., Heritage, G. L., Tooth, S., and Entwistle, N.: Morphodynamics of bedrock-influenced dryland rivers during extreme floods: insights from the Kruger National Park, South Africa, Geol. Soc. Am. Bull., in review, 2017.

Nanson, G. C.: Episodes of vertical accretion and catastrophic stripping: a model of disequilibrium flood-plain development, Geol. Soc. Am. Bull. 97, 1467–1475, 1986.

Rountree, M. W., Rogers, K. H., and Heritage, G. L.: Landscape state change in the semi-arid Sabie River, Kruger National Park, in response to flood and drought, South African Geographical Journal, 82, 173–181, 2000.

Tooth, S.: Changes in fluvial systems during the Quaternary, edited by: Knight, J. and Grab, S., Quaternary Environmental Change in Southern Africa: Physical and Human Dimensions, Cambridge University Press, 170–187, 2016.

van Niekerk, A. W., Heritage, G. L., and Moon, B. P.: River classification for management: the geomorphology of the Sabie River, South African Geographical Journal, 77, 68–76, 1995.

van Niekerk, A. W., Heritage, G. L., Broadhurst, L. J., and Moon, B. P.: Bedrock anastomosing channel systems: morphology and dynamics in the Sabie River, Mpumalanga Province, South Africa, edited by: Miller, A. J. and Gupta, A., Varieties of Fluvial Form, John Wiley and Sons, 219–247, 1999.

Proc. IAHS, 377, 57–66, 2018
https://doi.org/10.5194/piahs-377-57-2018
© Author(s) 2018. This work is distributed under
the Creative Commons Attribution 4.0 License.

Open Access PIAHS

Sedimentation and Its Impacts/Effects on River System and Reservoir Water Quality: case Study of Mazowe Catchment, Zimbabwe

Colleta Tundu[1], Michael James Tumbare[2], and Jean-Marie Kileshye Onema[3]

[1]Zimbabwe National Water Authority, P.O. Box Cy617 Causeway, Harare, Zimbabwe
[2]Department of Civil Engineering, University of Zimbabwe, P.O. Box MP167, Mt Pleasant, Harare, Zimbabwe
[3]WaterNet Secretariat, P.O. Box MP600, Mount Pleasant, Harare, Zimbabwe

Correspondence: Colleta Tundu (ctundu@gmail.com)

Received: 7 June 2017 – Accepted: 10 October 2017 – Published: 16 April 2018

Abstract. Sediment delivery into water sources and bodies results in the reduction of water quantity and quality, increasing costs of water purification whilst reducing the available water for various other uses. The paper gives an analysis of sedimentation in one of Zimbabwe's seven rivers, the Mazowe Catchment, and its impact on water quality. The Revised Universal Soil Loss Equation (RUSLE) model was used to compute soil lost from the catchment as a result of soil erosion. The model was used in conjunction with GIS remotely sensed data and limited ground observations. The estimated annual soil loss in the catchment indicates soil loss ranging from 0 to 65 $\mathrm{t\,ha\,yr^{-1}}$. Bathymetric survey at Chimhanda Dam showed that the capacity of the dam had reduced by 39 % as a result of sedimentation and the annual sediment deposition into Chimhanda Dam was estimated to be 330 t with a specific yield of 226 $\mathrm{t\,km^{-2}\,yr^{-1}}$. Relationship between selected water quality parameters, TSS, DO, NO_3, pH, TDS, turbidity and sediment yield for selected water sampling points and Chimhanda Dam was analyzed. It was established that there is a strong positive relationship between the sediment yield and the water quality parameters. Sediment yield showed high positive correlation with turbidity (0.63) and TDS (0.64). Water quality data from Chimhanda treatment plant water works revealed that the quality of water is deteriorating as a result of increase in sediment accumulation in the dam. The study concluded that sedimentation can affect the water quality of water sources.

1 Introduction

Sedimentation is a process whereby soil particles are eroded and transported by flowing water or other transporting media and deposited as layers of solid particles in water bodies such as reservoirs and rivers. It is a complex process that varies with watershed sediment yield, rate of transportation and mode of deposition (Ezugwu, 2013). Sediment deposition reduces the storage capacity and life span of reservoirs as well as river flows (Eroglu et al., 2010).

Sedimentation continues to be one of the most important threats to river eco-systems around the world. A study was done on the world's 145 major rivers with consistency long term sediment records and the results show that about 50 % of the rivers have statistically a significantly downward flow trend due to sedimentation (Walling and Fang, 2003). Sumi and Hirose (2009) reported that the global reservoir gross storage capacity is about 6000 $\mathrm{km^3}$ and annual reservoir sedimentation rates are about 31 $\mathrm{km^3}$ (0.52 %). This suggests that at this sedimentation rate, the global reservoir storage capacity will be reduced to 50 % by year 2100.

Studies on some dams in Zimbabwe show that reservoir capacities are being affected by sedimentation (Sawunyama et al., 2006; Dalu et al., 2013; Chitata et al., 2014).

Water is vital for all anthropogenic activities. Water bodies have been contaminated with various pollutants due to direct or indirect interference of men causing an adverse impact on human health and aquatic life (Lawson, 2011). The quality of water is getting vastly deteriorated due to improper land management and carelessness to the environment. Off late sediment transport in the water bodies has proved to be one of the major contributors to poor water quality. Due to land degradation and sheet erosion, the top soil is carried into the water bodies resulting in excess levels of turbidity. Silt and clay particles are primary carriers of adsorbed chemicals such as nitrogen and phosphorus.

This study uses the Mazowe Catchment area as a case study in analyzing river and reservoir sedimentation and its impact on water quality. Sedimentation in the rivers and reservoirs within the Mazowe Catchment area has become a major challenge for the policy makers as well as the water managers.

1.1 Study area

Mazowe Catchment is one of the seven water management catchments in Zimbabwe. It lies between 16.470° S (latitude) and 18.240° S (latitude) and between 30.680° E (longitude) and 33.000° E (longitude). The catchment has a total area of 38 005 km^2 and is in the northern part of the country. There are 30 major dams built along the main river and some of the tributaries. There are 13 functional gauging stations within the catchment. The catchment is composed of twenty-four (24) hydrological sub-zones. Rainfall for the Catchment averages 500 mm to 1200 mm yr^{-1} while the mean annual runoff is 131 mm with a coefficient of variation (CV) of 126 % (Zinwa, 2009). There are 17 water quality monitoring sampling points within the catchment. Six (6) sampling water points were selected for the study, these points were selected basing on the continuity of the available data meaning to say the points had less missing gaps.

Chimhanda Dam, in Lower Mazowe Sub-Catchment, was chosen for the assessment of the levels of sedimentation in resevoirs. Chimhanda Dam is located on the confluence of Runwa and Mwera Rivers, which are tributaries of Mazowe River. The dam is located in hydrological sub zone, DM1 with latitude 16°40′, and longitude 32°06′. The dam was completed in 1988 with a design capacity of 5.2×10^6 m^3 and covering a catchment area of 68.7 km^2, which mainly consists of communal lands.

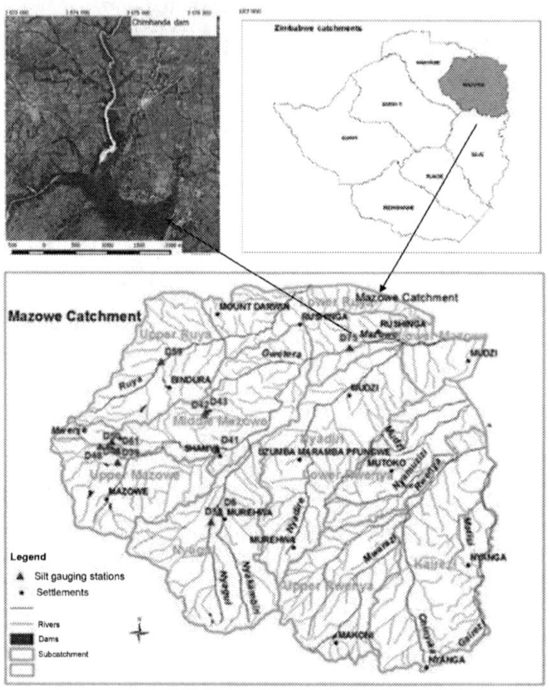

Figure 1. Mazowe Catchment and location of Chimhanda Dam.

2 Materials and methods

2.1 Data collection

Remote sensing images of the study area were downloaded from Landsat TM 4-5, LandSat LE7 and LandSat 8 scene from the website (https://glovis.usgs.gov/) for paths 168–170 and rows 71–73 for the period, 2000, 2005, 2008 and 2014. The study period was selected basing on the major changes in land use and land cover as a result of the Zimbabwe land reform programme which started in year 2000 and peaking during year 2003. The composite map was obtained by gluing and merging the tiles from the different scenes which was performed using Integrated Land and Water Information System (ILWIS) software. Normalized Differences Vegetation Index (NDVI) was extracted from the images and this was used to calculate the cover and management practice factor (C factor). SRTM DEM of 90m resolution was obtained from Earth Explorer website (http://EarthExplorer.com). The DEM was used to calculate the slope length and slope steepness factor (LS factor). The erodibility factor was obtained from the soil map of Zimbabwe that was downloaded from website (www.fao.org/geonetwork. Rainfall data was obtained from the Zimbabwe Meteorological Services Department and was used to come up with the erosivity factor. The grab sample method was used to obtain samples for sediment loads from the flow gauges for the 2014/2015 rainfall season. Historical data for sediment loads from flow gauging stations were obtained from the Zimbabwe National Water Authority (ZINWA). Historical water quality data for selected water quality monitoring stations was obtained from Environmental Management Agency (EMA).

Siltation historical data for selected reservoirs within the catchment was obtained from ZINWA. A bathymetric survey was done to assess the level of sedimentation of Chimhanda dam which lies within the Mazowe Catchment.

2.2 Calculation of RUSLE Factors

The Revised Universal Soil Erosion (RUSLE) model was used in this study to calculate the soil that was lost from the catchment. The model predicted the long term annual loss from the basin and is given by Eq. (1; Renard et al., 1997):

$$A = R \times K \times \mathrm{LS} \times C \times P \quad (1)$$

where: A = annual soil loss ($\mathrm{t\,ha\,yr^{-1}}$); R = rain erosivity factor ($\mathrm{MJ\,mm\,ha^{-1}\,h^{-1}}$); K = soil erodibility factor ($\mathrm{t\,ha\,h\,MJ^{-1}\,mm^{-1}}$); LS = slope length and slope steepness length (m); C = land cover and crop management; P = management practice.

2.2.1 Rain Erosivity Factor (*R* factor)

Rainfall data was processed into average annual rainfall. Rain erosivity was calculated from the rainfall point map using Eq. (2; Merritt et al., 2003):

$$R = 38.5 + 0.35 \times P \quad (2)$$

where: R = Rain Erosivity Factor ($\mathrm{Joule\,m^{-2}}$). P = Mean Annual Rainfall ($\mathrm{mm\,yr^{-1}}$).

2.2.2 Soil Erodibility Factor (*K* factor)

The soil erodibility factor (K Factor) was calculated using Eq. (3) with the parameters obtained from the soil map (Teh, 2011).

$$K = \frac{(1.0 \times 10^{-4}(12 - OM^{1.14} + 4.5(F-3) + 3.0(P-2))}{100} \quad (3)$$

where K = Soil Erodibility; M = (% fine sand +%fine sand)] × (100−%clay) O = % of organic matter; F = Soil Structure; P = Permeability.

2.2.3 Land Cover and Crop Management Factor (*C* factor)

The C factor was calculated using the Normalized Difference Vegetation Index (NDVI), which is a tool for assessing changes in vegetation cover (Pettorelli et al., 2005; Gusso et al., 2014). NDVI was calculated from the bands using Eq. (4; Deering, 1992).

$$\mathrm{NDVI} = \frac{(\mathrm{NIR} - \mathrm{Red})}{(\mathrm{NIR} + \mathrm{Red})} \quad (4)$$

where NIR = band 3 for landSat images 1 to 7 and band 4 for landSat 8. Red = band 4 for landSat 1 to 7 and band 5 for landSat 8.

The calculated NDVI was then used to calculate the C factor from Eq. (5)

$$C \text{ factor map} = 12708 \times \mathrm{NDVI} + 02585 \quad (5)$$

2.2.4 Management Practice Factor (*P* Factor)

The P factor reflects the control of conservation methods on soil loss. P values range from 0.01 to 1, with the value 0.01 being given to areas of maximum conservation support and the value 1 being given to areas with minimal or no conservation practices (Renard et al., 1997). The P values were derived from the land use map of the study area. Different values were assigned to each type of land use as guided by literature and the P factor map was produced (Jang et al., 1996).

2.3 Quantifying the sediment yield in the Catchment

The estimated soil loss was calculated from Eq. (6).

$$A = R \times K \times \mathrm{LS} \times C \times P \quad (6)$$

where A = Average Soil Loss ($\mathrm{t\,ha\,yr^{-1}}$); R = R factor map; K = K factor map; LS = LS factor map; C = C factor map; P = P factor.

Soil loss from the catchment cannot be taken as sediment contribution to a river flow system since it does not account for deposition that occurs along the path (de Vente et al., 2011). Therefore the estimated soil loss was multiplied by the sediment delivery ratio (SDR) to obtain the sediment yield of the catchment. Sediment delivery ratios represent the fraction of the total soil loss that is washed into rivers and was calculated from using Eq. (7; USDA, 1972).

$$\mathrm{SDR} = 05656\mathrm{CA}^{-0.11} \quad (7)$$

where SDR = Sediment Delivery Ratio; CA = Watershed Area, $\mathrm{km^2}$.

After determination of the sediment delivery ratio, the average sediment yield was determined using Eq. (8), by Wischmeier and Smith (1978)

$$\mathrm{SR} = \mathrm{SDR} \times A \quad (8)$$

where: SR = Sediment yield ($\mathrm{t\,ha\,yr^{-1}}$); SDR = Sediment delivery ratio; A = Average soil loss ($\mathrm{t\,ha\,yr^{-1}}$).

2.4 Sediment yield from field measurements

Seven out of the thirteen functional ZINWA gauging stations were selected for collection of sediment concentration samples, (Fig. 2). Water samples were collected from various flow gauging stations for the period November 2014 to March 2015 using grab sampling method. An average of 20 samples was collected from each gauging station. The samples were analyzed in a laboratory using the weighing and filtration method inorder to determine the sediment concentration in $\mathrm{mg\,L^{-1}}$. The sediment load was determined by multiplying the sediment load at a particular gauge by the area of influence.

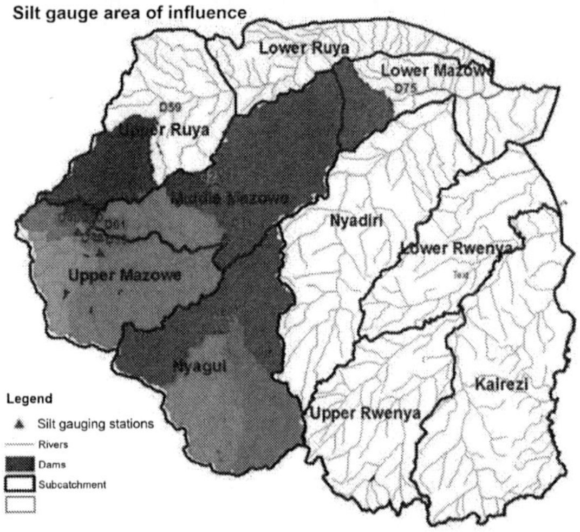

Figure 2. Gauges in the catchment and areas of influence.

2.5 Assessing the current sedimentation levels of Chimhanda Dam

A bathymetric survey was carried out at Chimhanda dam to determine the siltation level of the dam. The survey was conducted using a SonTek River Surveyor system and a theodolite to come up with a basin survey map. The plotted map was digitized to get the surface area between contours. The area was then multiplied by the contour interval to get the volume of each contour using Simpson's formula.

$$V_{\text{contour}} = \frac{A_1 + (A_1 \times A_2)^{1/2} + A_2}{3} \quad (9)$$

where V_{contour} = contour Volume (m^3); A_1 = Area_1 (m^2); A_2 = Area_2 (m^2).

The calculated volumes for each contour were accumulated to get the new capacity of the dam.

2.6 Water quality trend from sedimentation

Water quality historical data from six selected water sampling points were obtained from the Environmental Management Agency records. The points were selected basing on the consistency of the data and less gaps. The trend for the selected water quality parameters, TSS, DO, NO_3, pH, TDS and turbidity for the corresponding predicted sediment yield years were analysed. The water quality parameters were compared with the Environmental Management Agency, Effluent and Solid Waste Disposal Regulations Statutory Instrument number 6 of 2007 (EMA, 2007) shown on Table 1. Water quality data for Chimhanda water supply treatment plant from the Zimbabwe National Water Authority was also related to the change in level of sedimentation of Chimhanda Dam.

2.7 Investigation the relationship between sedimentation and water quality

Selected water quality parameters were correlated with the corresponding sediment yield to determine the relationship between sediment yield and water quality. Pearson Correlation was used to estimate the strength of relationship between sediment yield and some physical water quality.

3 Results and discussions

3.1 Results of the RUSLE factors

The erosivity map (R factor) depicts rainfall energy in the various areas within the catchment. The rainfall erosivity ranges between 200 and 500 ($\text{MJ mm ha}^{-1}\,\text{h}^{-1}\,\text{yr}$). A greater part of the catchment is averaging a rainfall erosivity value of $276\,\text{MJ mm ha}^{-1}\,\text{h}^{-1}\,\text{yr}$. Highest rainfall erosivity values are in the Kairezi sub-catchment of the study area. Erodability (K factor) values range from 0 to $0.5\,\text{t ha}^{-1}\text{yr}^{-1}\,\text{MJ}^{-1}\,\text{mm}^{-1}$. Some parts in the south west of the catchment have high values of erodibility, the highest K value is dominated by very fine sand with silt particles which give rise to higher soil erodibility (Kamaludin et al., 2013). Highest slope length steepness (LS) values of 16 m were found in the eastern part of the catchment. The northern part of the catchment experiences average LS values of 7 m.

Cover Management and Practice (C factor) ranges from -1 to 1. The C factor is depicted by NDVI which is a function of photosynthesis. The C factor is high in the eastern highlands area, some parts of Kairezi sub-catchment and along the Mufurudzi area. This is possible because the Kairezi area experiences high rainfall while Mufurudzi being a game park, has more vegetation. High C values were observed with high vegetation cover (Pettorelli et al., 2005). The P value of 1 was observed in Lower Mazowe and Upper Rwenya subcatchment areas respectively. Areas like Mufurudzi and parts of Kairezi experienced an average P value of 0.8.

3.2 Soil loss from the catchment

The average annual soil loss for the different years are shown on Fig. 3 and the temporal variation of actual soil loss is tabulated on Table 2.

The estimated soil loss from the catchment ranges from 0 to $203\,\text{t ha yr}^{-1}$. The estimated average soil loss is $54\,\text{t ha yr}^{-1}$. The highest soil loss is being experienced in Middle Mazowe and Nyagui sub-catchments. High soil loss in Middle Mazowe sub-catchment can be associated with high gold panning activities in the Mazowe valley whilst in Nyagui sub-catchment, this can be associated with high crop farming activities in the area. When the soil is made loose, its structure is altered hence increase in erodibility.

Table 1. Effluent and solid waste disposal regulations.

PARAMETER	BANDS Blue		Green	Yellow	Red	Test Methods
	Sensitive	Normal				
DO % saturation	> 75	> 60	> 50	> 30	> 15	SAZS 573
Nitrates NO_3 mg L^{-1}	< 10	< 10	< 20	< 30	< 50	
PH	6.0–7.5	6.0–9.0	5–6, 9–10	4–5, 10–12	0–4, 12–14	SAZS 483
TDS	< 100	< 500	< 1500	< 2000	< 3000	SAZS 576,
Turbidity (NTU)	< 5	< 5				SAZS 478
TSS mg L^{-1}	< 10	< 25	< 50	< 100	< 150	SAZS 478

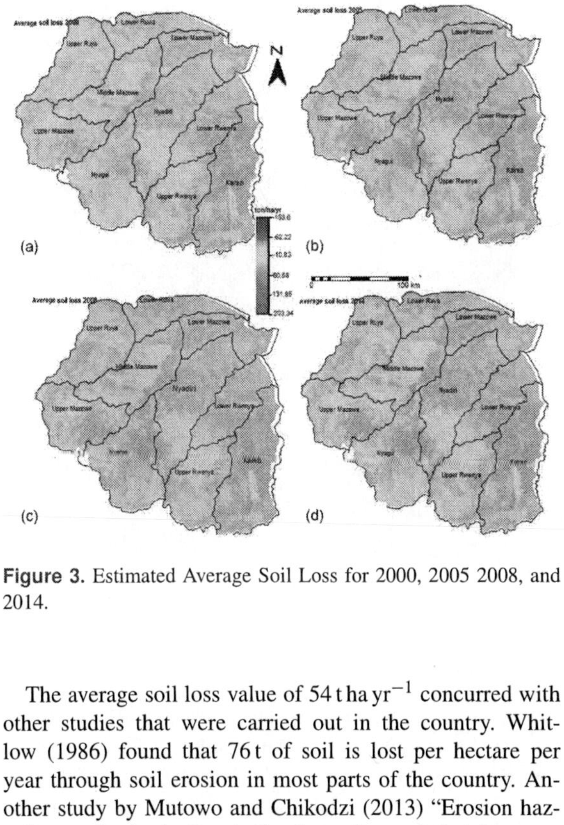

Figure 3. Estimated Average Soil Loss for 2000, 2005 2008, and 2014.

Table 2. Temporal Variation of Estimated Soil Loss.

Year	Average Soil Loss (t ha yr^{-1})
2000	54
2005	65
2008	36
2014	62

The average soil loss value of 54 t ha yr^{-1} concurred with other studies that were carried out in the country. Whitlow (1986) found that 76 t of soil is lost per hectare per year through soil erosion in most parts of the country. Another study by Mutowo and Chikodzi (2013) "Erosion hazards mapping in the Runde Catchment", concluded that most of the areas in the catchment fall in the category range of 0–50 t ha yr^{-1}. Makwara and Gamira (2012) reported that the most serious type of erosion being sheet erosion, is estimated to remove an average 50 t ha yr^{-1} from Zimbabwe's communal lands.

There is no uniform trend in the soil loss over the study period. In year 2000, the soil loss was 54t ha yr^{-1}. The increase of soil loss in 2005 from the 2000 figure can be explained by the land reform programme which started in year 2000 and was at its peak around 2003. There was a lot of deforestation as new farmers were clearing land for agricultural purposes (Mambo and Archer, 2007). Some areas which were meant for animal rearing and forest were also converted to crop agricultural areas. In 2008 the soil loss rate decreased to 36 t ha yr^{-1}. The reduction can be attributed to a number of factors. Year 2008 experienced low erosivity as a result of low rainfall, which could also lead to low annual soil loss. The economy of Zimbabwe was almost at a stand-still during the period 2007 to 2009, with an inflation rate record of over 231 million percent in July 2008 (Hanke and Kwok, 2009). As a result there was less farming activities during that period.

The soil loss rate almost doubled in year 2014 as compared to 2008. This can be explained by the increase in the number of small scale gold miners in the catchment and alluvial gold panning in streams and rivers. Mining within the catchment is not only limited to the river beds and banks, miners are also targeting the inland areas of the catchment such as the Mazowe Valley and Mufurudzi Game Park and as a result, soil erodibility is also increasing. Increase in rate of unemployment caused by low capital investment and continuing closure of industries in Zimbabwe has resulted in the population resorting to other sources of income such as illegal gold mining and alluvial gold panning. A study that was carried out in the Lower Manyame sub – catchment along Dande River, on the analysis of the implications of cross- sectional coordination of the management of gold panning activities

Table 3. Measured sediment Load from flow gauges for the period December 2014 to March 2015.

Silt Gauge	Name or River	Area of Influence (km^2)	Sediment load ($t\,km^{-2}$)	Load (t)
D75	Mazowe	20 380	1.4	28536
D50	Nyamasanga	13	0.24	3
D42	Mupfurudzi	163	1.92	313.56
D41	Mazowe	3300	0.86	2836
D6	Shavanhowe	1166	0.51	589.65
D58	Nyagui	98.3	24.25	142.93
D48	Mwenje	399	0.36	33.4

Table 4. Summarized Results from the silt survey.

	1988	2003	2015
Capacity at FSL ($\times 10^6\,m^3$)	5.2	3.470	3.194
Loss Of Storage (%)	0	33.3	38.6
Dam Surface Area (Ha)	93.00	74.00	64.04

and its impacts (Zwane et al., 2006), identified river bed gold panning activities as a cause of degradation of river channels and banks as well as accelerated erosion and siltation in many areas of Zimbabwe.

3.2.1 Sediment Yield

Figure 4 shows the sediment yield distribution map. Sediment yield ranges from less than $1\,t\,ha\,yr^{-1}$ to a maximum of $24.5t\,ha\,yr^{-1}$ with an average value of $6.04\,t\,ha\,yr^{-1}$ over the catchment area. The highest sediment yield is in the Middle Mazowe sub catchment area with an average of $6.72\,t\,ha\,yr^{-1}$ and a maximum of $24.5t\,ha\,yr^{-1}$. The high value can be associated with high gold panning activities in the Mazowe valley. The lowest value is in the Lower Ruya sub-catchment with an average of $4\,t\,ha\,yr^{-1}$. The results are almost in line with a study by Rooseboom and Engineers (1992) that estimated the average sediment yield in the nine defined sediment regions in Southern Africa to vary between 30 and $330\,t\,ha\,yr^{-1}$ with an average of $15\,t\,ha\,yr^{-1}$. A study which was also carried out in the catchment area of Pahang River basin in Malasyia gave a sediment yield ranging from 0–$13.79\,t\,ha\,yr^{-1}$ (Kamaludin et al., 2013). The total sediment yield for the catchment was calculated to be 6.3 million $t\,yr^{-1}$.

3.2.2 Measured Sediment Loads

Results from the measured sediment loads were highest at D75 which is along Mazowe River, in the Zambezi valley. D41, located in Middle Mazowe sub catchment recorded a relatively high sediment load of 2836 t. D58 and D6 are both in Nyagui Sub-catchment recorded 142.93 and 589.65 t respectively. D50, which is in Upper Mazowe sub catchment, recorded a low figure of 3 t. The results are concurring with those from RUSLE model, where high sediment yields were recorded from Middle Mazowe and Nyagui sub catchments respectively.

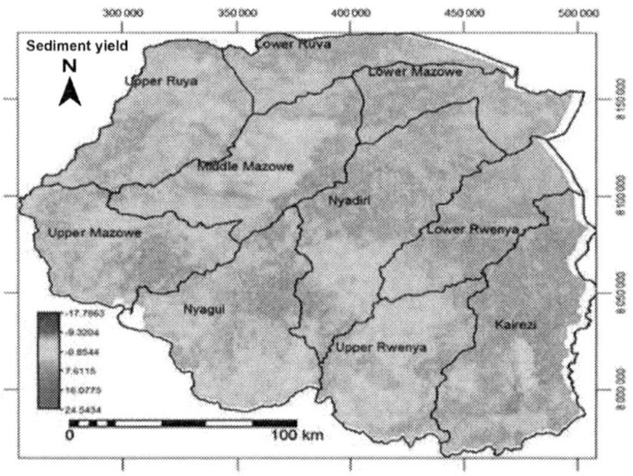

Figure 4. Sediment Yield within the Catchment.

Table 5. Results for the correlation between sediment yield and water quality parameters.

		Sediment	Ave Turbidity	Average TSS
Sediment	Pearson Correlation	1	.634	.647
	Sig. (2-tailed)		.366	.353
	N		28	28

3.3 Sedimentation in Chimhanda Dam from Bathymetric Survey

Table 4 summarizes the results of the silt survey for Chimhanda Dam.

The capacity of the dam decreased from the original 5.200×10^6 to 3.470×10^6 and $3.194 \times 10^6\,m^3$ in years 2003 and 2015 respectively. There was 33.3 % loss in stor-

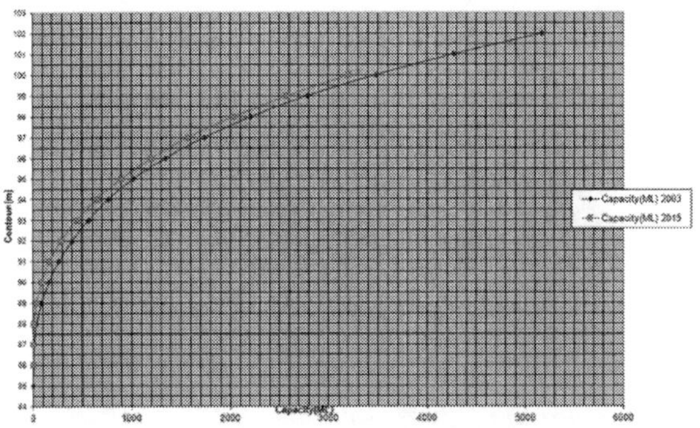

Figure 5. Chimhanda Dam Capacity comparison over the years.

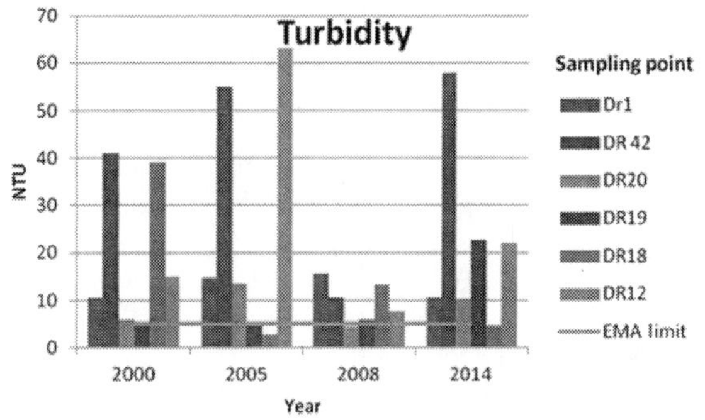

Figure 6. Turbidity variation among sampling points.

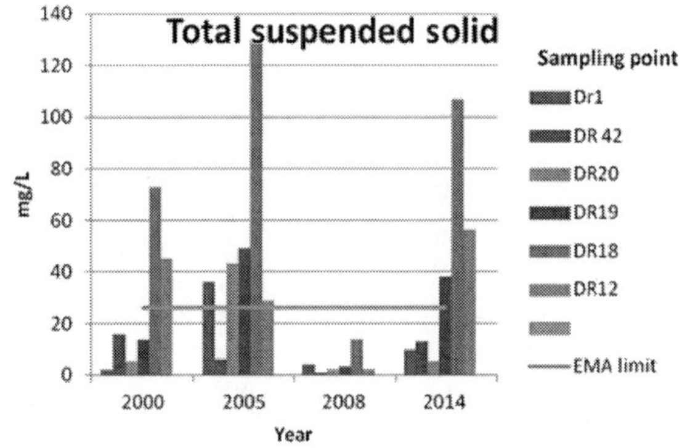

Figure 7. Total Suspended Solids variation among sampling points.

age up to 2003 and a storage loss of 38.6 % up to 2015 in reference to the design capacity. Figure 5 shows reservoir capacity changes over the years.

Results from the 10 samples of river inflows collected gave an average sediment concentration of $39\,mg\,L^{-1}$. There was a decrease in trend in the sediment concentration as the rainfall season was progressing. The sediment accumulation in the reservoir was found to be $330\,t\,yr^{-1}$ and the specific sediment yield was $226\,t\,km^{-2}\,yr^{-1}$. The usefulness of Chimhanda Dam has reduced from the initial design life of 50 to 37 years. A study by Godwin et al. (2011) on Chesa Causeway Weir in the same catchment gave a specific yield of $510\,t\,km^2\,yr^{-1}$. The usefulness of the dam was reduced from the initial design life of 50 to 25 years.

Generally the results support the existing literature that most of the small to medium dams are being affected by high rates of sedimentation in Zimbabwe (Sawunyama et al., 2006; Godwin et al., 2011; Dalu et al., 2013; Chitata et al., 2014).

3.4 Water Quality Trend from Sedimentation

Figures 6 to 10 show the results for the historically measured water parameters for the years 2000, 2008, 2010 and 2014.

3.4.1 Turbidity

Results for the years did not meet the permissible turbidity of 5 NTU. The highest turbidity values were recorded in 2000 and 2005 respectively. High values in 2005 corresponds to the high soil loss of $65\,t\,ha\,yr^{-1}$ that was recorded during that particular year. The high values in 2000 can be explained by high rainfalls that were experienced during that particular year as compared with the other years (Met, 2005). Low turbidity values were recorded during the year 2008, the same year recorded the lowest soil loss of $36\,t\,ha\,yr^{-1}$. High Turbidity levels are associated with poor water quality, (Adekunle et al., 2007). WHO (2008) highligted that high turbidity waters can facilitate the formation of nuclei, where gastrointestinal disease pathogens can attach. If that water is consumed improperly treated, it can cause diseases. High turbidity leves also render the treatment of water expensive.

3.4.2 Total Suspended Solids

2008 values were below the recommended limit of $15\,mg\,L^{-1}$ as shown on Fig. 7. High values were recorded during year 2000, 2005 and 2014. The high values are in agreement with the high soil loss that were recorded in those particular years. Total suspended solids are closely related to sedimentation (Chapman, 1996). When these suspended particles settles at the bottom of a water body they become sediments. Suspended solids consist of inorganic and organic fractions. Part of the organic fractions is bacterial and that might be detrimental to human health if the water is consumed without adequate treatment (Hoko, 2008).

3.4.3 Dissolved Oxygen (DO)

Dissolved Oxygen was above the lower limit for EMA at all the stations for all the years. Year 2000 and 2005 recorded % saturation of 125 and 100 respectively. Oxygen levels can be higher due to excess aquatic plants in the water which

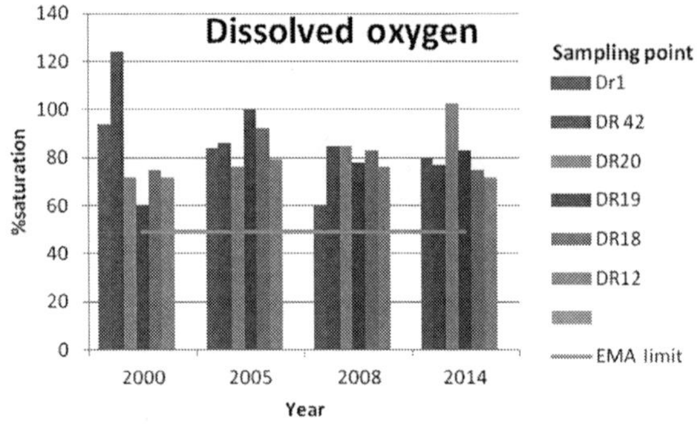

Figure 8. Dissolved Oxygen variation among sampling points.

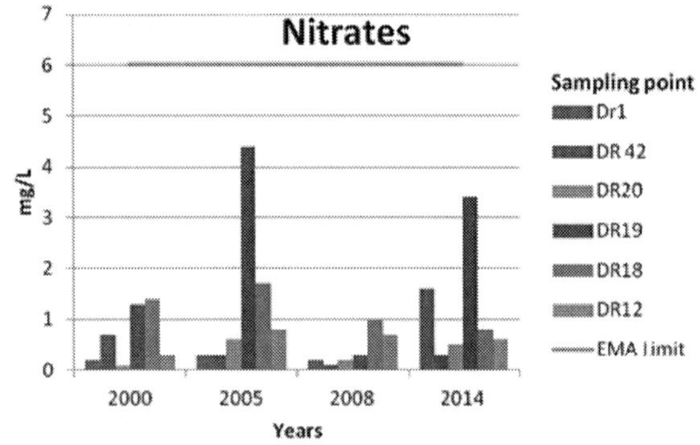

Figure 9. Nitrates variation along sampling points.

produce oxygen. Most living organisms require oxygen for their basic metabolic processes.

3.4.4 Nitrates (NO_3)

Results for nitrates for the years under consideration were all below the recommended EMA levels of $6\,mg\,L^{-1}$. Nitrates are washed by runoff from agricultural lands into water bodies. High concentration of nitrates in water bodies cause eutrophication, resulting in competition for oxygen by aquatic organisms and high water treatment costs (Dike et al., 2010).

3.4.5 PH

pH values for all the years were within the normal limits category according to EMA upper and lower limits of 6 and 9 units respectively. Year 2005 had one sampling point that had a value slightly lower than the recommended. Bhadja and Vaghela (2013) indicated that activities in the water shed such as increased leaching of soils, soil erosion and heavy precipitation affects the pH levels downstream. pH levels as high as 9.0 and as low as 5 affects the life span of aquatic organisms (Addo et al., 2011). Low pH values do not have direct effect to human health, but can have indirectly effect

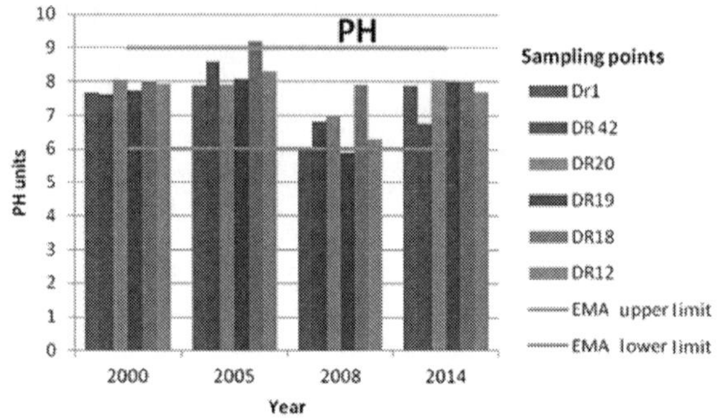

Figure 10. pH variations along sampling points.

since it causes leaching of metal irons such as copper, zinc and lead into the water (WHO, 2008).

3.5 Relationship between sedimentation and river water quality

Pearson correlation was used to determine the relationship between two water quality parameters and sediment yield. The parameters are related to sediment yield and these are turbidity and suspended solids, (Bhadja and Vaghela, 2013). The results are shown on Table 5.

There is a close positive relationship between turbidity and sediment yield of 0.63. Total Suspended Solids have a positive relation of 0.64. The results show that the two water quality parameters can be indicators of sedimentation. Higher values of total suspended solids and turbidity indicate higher values of sediment yield in water bodies.

3.6 Variation in reservoir water quality

The variation in two selected reservoir water quality indicators, turbidity and pH was determined. A significant variation in turbidity was noted for the years 1988 to 2014 with a p-value of 0.02, which is less than 0.05. The mean value of turbidity for 1988, 2003 and 2014 are 15.6, 34.4 and 41.7 NTU respectively. The values show an increasing trend in turbidity which may be as a result of anthropogenic activities occurring upstream of the dam which has resulted in increased soil erosion hence more sediment on the reservoir. pH results showed that there is no significant change between 1988 and 2014 (p-value > 0.05). This might be because pH is a chemical parameter that is determined by the type of sediments that are carried into the river not the sediment load.

4 Conclusions and recommendations

4.1 Conclusion

The objective of the study was to analyze sedimentation and its impact on river system and reservoir water quality in Mazowe Catchment, Zimbabwe. The study's major findings are,

The catchment is generally experiencing moderate soil loss, however there are some pockets under high and extreme soil loss which gave rise to siltation and water quality deterioration of available water bodies.

There is reduction in capacity of reservoirs within the catchment due to sedimentation. Chimhanda Dam capacity has been reduced by 38.6 % from the initial design capacity as of April 2015, reducing the usefulness of the dam to 37 years from the initial design life of 50 years due to siltation.

Sediment deposition in rivers affects water quality and water quality parameters can be used as an indicator for sedimentation.

4.2 Recommendations

In order to curb erosion and sedimentation in rivers and reservoirs, there is need to develop and implement an integrated water resources management plan by all stakeholders.

In order to maintain the lifespan of Chimhanda dam, best land management practices such as contour ploughing and re-forestation should be employed around the catchment area of the Dam and this is an area that could be considered for further research.

Data availability. The main data used in the research is rainfall data that was obtained from the meteorological department, water quality data obtained from Environmental management Agency as well as sediment Data from Zinwa. The data can be obtained from these departments.

Competing interests. The authors declare that they have no conflict of interest.

Special issue statement. This article is part of the special issue "Water quality and sediment transport issues in surface water". It is a result of the IAHS Scientific Assembly 2017, Port Elizabeth, South Africa, 10–14 July 2017.

Acknowledgements. This paper contains part of research results from the MSc Thesis of the corresponding author, Colleta Tundu, submitted to the University of Zimbabwe. Special thanks go to Faith Chivava for her assistance in GIS. ZINWA, EMA and the Zimbabwe Meteorological Services Department is appreciated for providing data.

Edited by: Akhilendra B. Gupta
Reviewed by: Rajendra Prasad and one anonymous referee

References

Addo, M. A., Okley, G. M., Affum, H.A., Acquah, S., Gbadago, J. K., Senu, J. K., and Botwe, B. O.: Water Quality and level of some heavy metals in water and sediments of Kpeshie Lagoon, La-Accra, Ghana, Research Journal of Environmental and Earth Sciences, 3, 487–497, 2011.

Adekunle, I. M., Adetunji, M. T., Gbadebo, A. M., and Banjoko, O. B.: Assessment of groundwater quality in a typical rural settlement in Southwest Nigeria, Int. J. Env. Res. Pub. He., 4, 307–318, 2007.

Bhadja, P. and Vaghela, A. : Assessment of Physico-Chemical parameters and water quality Index of reservoir water, International journal of plant animal and environmental sciences, 3, 89–95, 2013.

Chapman, D.: Water quality assessments – A guide to use of Biota, sediments and water in Environmental monitoring, 2nd Edn., Chapman and Hall, London, 191–193, 1996.

Chitata, T., Mugabe, F. T., and Kashaigili, J. J.: Estimation of Small Reservoir Sedimentation in Semi-Arid Southern Zimbabwe, Journal of Water Resource and Protection, 6, 48909, 12–14, 2014.

Dalu, T., Tambara, E. M., Clegg, B., Chari, L. D., and Nhiwatiwa, T.: Modeling sedimentation rates of Malilangwe reservoir in the south-eastern lowveld of Zimbabwe, Applied Water Science, 3, 133–144, 2013.

Dearing, J.: Sediment yields and sources in a welsh upland lake-catchment during the past 800 years, Earth Surf. Proc. Land., 17, 1–22, 1992.

de Vente, J., Verduyn, R., Verstraeten, G., Vanmaercke, M., and Poesen, J.: Factors controlling sediment yield at the catchment scale in NW Mediterranean geoecosystems, J. Soil. Sediment., 11, 690–707, 2011.

Dike, L., Keramat A., and Amirkolaie S.: Environmental impact of nutrient discharged by aquaculture waste water on the Haraz river, Journal of Agriculture and Biological Science, 3, 275–279, 2010.

EMA: Effluent and Solid Waste Disposal Regulations SI 6, of 2007, government printers of Zimbabwe, 2007.

Eroğlu, H., Çakır, G., Sivrikaya, F., and Akay, A. E.: Using high resolution images and elevation data in classifying erosion risks of bare soil areas in the Hatila Valley Natural Protected Area, Turkey, Stoch. Env. Res. Risk A., 24, 699–704, 2010.

Ezugwu, C.: Sediment Deposition in Nigeria Reservoirs: Impacts and Control Measures, Innovative Systems Design and Engineering, 4, 54–62, 2013.

Godwin, A. M., Gabriel, S., Hodson, M., and Wellington, D.: Sedimentation impacts on reservoir as a result of land use on a selected catchment in Zimbabwe, International Journal of Engineering Science & Technology, 3, 2011.

Gusso, A. G. A., Veronez, M. R., Robinson, F., Roani, V., and Da Silva, R. C.: Evaluating the thermal spatial distribution signature for environmental management and vegetation health monitoring, International Journal of Advanced Remote Sensing and GIS, 3, 433–445, 2014.

Hanke, S. H. and Kwok, A. K.: On the measurement of Zimbabwe's hyperinflation, Cato J., 29, 353–364, 2009.

Hoko, Z.: An assessment of quality of water from boreholes in Bindura District, Zimbabwe, Phys. Chem. Earth, 33, 824–828, 2008.

Jang, C., Shin, Y., Kum, D., Kim, R., Yang, J. E., Kim, S. C., Hwang, S. I., Lim, K. J., Yoon, J.-K., and Park, Y. S.: Assessment of soil loss in South Korea based on land-cover type, Stoch. Env. Res. Risk A., 1–15, 1996.

Kamaludin, H., Lihan, T., Ali Rahman, Z., Mustapha, M. A., Idris, W. M. R., and Rahim, S. A.: Integration of remote sensing, RUSLE and GIS to model potential soil loss and sediment yield (SY), Hydrol. Earth Syst. Sci. Discuss., https://doi.org/10.5194/hessd-10-4567-2013, 2013.

Lawson, E. O.: Physico- chemical parameters and heavy metal contents of water from the mangroves swamps of logos lagon, logos, Nigeria, Advances in biological research, 5, 8–21, 2011.

Makwara, E. and Gamira, D.: About to Lose all the Soil in Zaka's Ward 5, Zimbabwe: Rewards of Unsustainable Land Use, European Journal of Sustainable Development, 1, 457–476, 2012.

Mambo, J. and Archer, E.: An assessment of land degradation in the Save catchment of Zimbabwe, Area, 39, 380–391, 2007.

Merritt, W. S., Letcher, R. A., and Jakeman, A. J.: A review of erosion and sediment transport models, Env. Modell. Softw., 18, 761–799, 2003.

Met: Meteorological data, available at: https://www.wmo.int/datastat/wmodata_en.htmll, last access: 3 March 2015.

Mutowo, G. and Chikodzi, D.: Erosion Hazard Mapping in the Runde Catchment: Implications for Water Resources Management, Journal of Geosciences and Geomatics, 1, 22–28, 2013.

Pettorelli, N., Vik, J. O., Mysterud, A., Gaillard, J.-M., Tucker, C. J., and Stenseth, N. C.: Using the satellite-derived NDVI to assess ecological responses to environmental change, Trends Ecol. Evol., 20, 503–510, 2005.

Renard, K. G., Foster, G. R., Weesies, G. A., McCool, D., and Yoder, D.: Predicting soil erosion by water: a guide to conservation planning with the revised universal soil loss equation (RUSLE), Agriculture Handbook, Washington, 703, 1997.

Rooseboom, A. and Engineers, S. B. C.: Sediment transport in rivers and reservoirs: a Southern African perspective, Water Research Commission, 297/1/92:489, 1992.

Sawunyama, T., Senzanje, A., and Mhizha, A.: Estimation of small reservoir storage capacities in Limpopo River Basin using geographical information systems (GIS) and remotely sensed surface areas: Case of Mzingwane catchment, Phys. Chem. Earth, 31, 935–943, 2006.

Sumi, T. and Hirose, T.: Accumulation of sediment in reservoirs, Water storage, transport and distribution, UNESCO-IHE and EOLSS Publishers Co. Ltd., Paris, France, 224–252, 2009.

Teh, S. H.: Soil erosion modeling using RUSLE and GIS on Cameron highlands, Malaysia for hydropower development, MS, The School for Renewable Energy Science, Iceland, 41 pp.,2011.

USDA: Procedure for computing sheet and rill equation on project areas, Technical Release No. 51 (Geology), U.S. Government Printing Office, Washington, D.C., 1972.

Walling, D. and Fang, D.: Recent trends in the suspended sediment loads of the world's rivers, Global Planet. change, 39, 111–126, 2003.

Whitlow, R.: Mapping erosion in Zimbabwe: a methodology for rapid survey using aerial photographs, Appl. Geogr., 6, 149–162, 1986.

WHO: Guidelines for drinking water quality: Third edition incorporating the first and second addenda volume 1 additions, 2008.

Wischmeier, W. and Smith, D.: Predicting rainfall erosion losses. Agricultural Handbook 537, Agricultural Research Service, United States Department of Agriculture, 1978.

Zinwa, D. R.: Assessment of Surface Water Resources of Zimbabwe and Guidelines for Planning, in: Management, ZINWA Research and Data Department, Ministry of Water Resources Development and Management Harare, Zimbabwe, Harare, 59 pp., 2009.

Zwane, N., Love, D., Hoko, Z., and Shoko, D.: Managing the impact of gold panning activities within the context of integrated water resources management planning in the Lower Manyame Sub-Catchment, Zambezi Basin, Zimbabwe, Phys. Chem. Earth, 31, 848–856, 2006.

Proc. IAHS, 377, 67–76, 2018
https://doi.org/10.5194/piahs-377-67-2018
© Author(s) 2018. This work is distributed under
the Creative Commons Attribution 4.0 License.

Etude de l'impact des barrages sur la réduction des transports sédimentaires jusqu'à la mer par approche paléohydrologique dans la basse vallée de la Medjerda

Fatma Kotti[1], Laurent Dezileau[3], Gil Mahé[2], Hamadi Habaieb[1], Malik Bentkaya[4], Claudine Dieulin[2], and Oula Amrouni[5]

[1]Institut National Agronomique de Tunis, 43, Av. Charles Nicolle 1082 Tunis-Mahrajène, Tunisie
[2]UMR HydroSciences Montpellier/IRD, Montpellier, France
[3]Géoscience Montpellier, Université de Montpellier, Montpellier, France
[4]Centre National des Sciences et Technologies Nucléaires, Sidi Thabet, Tunisie
[5]Institut National des Sciences et Technologies de la Mer, Salammbô, Tunisie

Correspondence: Fatma Kotti (f.kotti@yahoo.fr)

Received: 15 June 2017 – Accepted: 4 October 2017 – Published: 16 April 2018

Résumé. The sedimentary contributions of the Medjerda to the coastal zone are poorly measured, and there is no chronicle of observations. In this context, the sediment monitoring appears indispensable for the quantification of sediment transport at the outlet. This study focuses on the largest watershed in Tunisia, the Wadi Medjerda (23 600 km^2). The main objective of this work is to assess the reduction of sediment transport following anthropogenic intensification on the basin, especially since the construction of many large dams. In order to collect information on actual deposits over several years, the paleo-hydrological approach was applied through the study of sediment cores sampled in the low valley meanders on alluvial terraces, after the last dam (Sidi Salem, the largest water storage capacity over the basin), but before the estuary to avoid marine influence and near a hydrological station (Jdaida). The sedimentary deposits of the river provide key information on the past sedimentary inputs. A visible succession of sedimentary layers corresponding to the deposits of successive floods on the study site has been determined and the history of the sedimentary contributions of the Medjerda is reconstructed by this approach. The thickest layers of sedimentary deposits are related to exceptional events. They are mainly concentrated on the lower part of the core and are mainly composed of sands. The first 1.2 m of the core from the bottom upward relates to 10 years of river discharges, as can be determined from the ^{137}Cs datation. The next upward 1.05 m of core relates to the following 20 years of discharges, up to 1981, date of the construction of the Sidi Salem dam, and is composed of a mix of sand, silts and clays. The last 75 cm of core near the surface is only composed of clays with thin silt bands, and relates to a period of 32 years. We thus observe that there is no more sand deposits in the river bed since the construction of the Sidi Salem dam. The deficit of sediment supply to the sea is viewed as a major factor to be taken into account for better understanding of the dynamics of coastal areas in the context of global climate change.

1 Introduction

Dans le contexte actuel des changements climatiques, le suivi du transport sédimentaire dans les rivières est important pour mettre l'accent sur les problèmes environnementaux souvent reliés à des changements observés dans le cycle hydrologique : modification de la distribution des précipitations, du ruissellement et de l'intensité des extrêmes hydrologiques (Bates et al., 2008 ; Hungtinton, 2006 ; Labat et al., 2004 ; Solomon, 2007). Le suivi du transport sédimentaire permet également d'évaluer et de quantifier les transferts de matières des continents vers les océans. Cependant, le problème du nombre réduit de données relatives aux transports solides a toujours freiné les investigations. Les mesures de transport solide effectuées en Tunisie et les études ponctuelles détaillées restent insuffisantes pour traiter ce sujet (Ghorbel et Claude, 1982). Le défi consiste donc à utiliser des méthodes de mesures simples pour combler le manque de connaissances. La présente étude s'intéresse au fleuve Medjerda en Tunisie, afin d'étudier en détail l'évolution des transports solides à l'exutoire de son bassin versant. L'Oued Medjerda est devenu un fleuve presque totalement contrôlé depuis l'intensification des aménagements sur le bassin au début du XXème siècle, pour atténuer les effets des crues dans la basse vallée très intensément exploitée. Le transport de matières solides a été considérablement réduit dans sa basse vallée, en particulier depuis l'ouverture du barrage Sidi Salem sur le cours principal de la Medjerda, en 1981, responsable d'une modification majeure du régime hydrologique du fleuve en aval (Zahar et al., 2008). L'alluvionnement des retenues des aménagements constitue un problème récent et majeur (Ben Mamou et Louati, 2007). Or, l'histoire des apports sédimentaires dans cette zone d'étude n'est pas connue, ce qui entraîne deux difficultés :

- comment évaluer l'apport solide à la mer si l'histoire des apports sédimentaires n'est pas connue ?
- comment identifier les différents facteurs contrôlant l'apport sédimentaire par la Medjerda ?

La paléohydrologie peut en partie répondre à ces questions. On tentera dans ce cas d'établir un historique de la sédimentation dans le bief aval du fleuve Medjerda, avant la mer, et de relier cet historique avec l'histoire des débits, l'évolution climatique et la construction des aménagements. La paléohydrologie est la reconstruction de l'ampleur et de la fréquence des inondations récentes et passées en utilisant des archives sédimentaires (Baker et al., 2002). Notre travail va se concentrer principalement sur la Medjerda, près de la station hydrologique de Jdaida située dans la basse vallée mais suffisamment loin de la mer pour ne pas être impactée par la marée, zone d'étude pour laquelle il existe une quantité importante de données historiques, particulièrement sur les débits des 70 dernières années. Notre approche consiste à prélever des carottes de sédiment dans un méandre où les dépôts sont réguliers et peu turbulents, afin de récolter des informations sur les dépôts réels sur plusieurs décennies. Nous avons utilisé des techniques d'analyse stratigraphique et de datation des dépôts (^{137}Cs, ^{210}Pb, géochimie) pour évaluer la hauteur atteinte par les eaux lors de différents évènements d'inondation, et proposer une chronologie des crues de la Medjerda. L'analyse pluridisciplinaire de ces archives sédimentaires permet de fournir de précieuses informations sur l'évolution climatique des dernières décennies, les événements extrêmes et enfin sur les effets des actions anthropiques passées et récentes (Vilanova et al., 2006 ; Selby et Smith, 2007).

2 Matériel et méthode

2.1 Site d'étude du bassin versant de la Medjerda

Le fleuve Medjerda est le plus long de Tunisie, avec 484 km dont 350 km en Tunisie (Ben Mansoura et al., 2001). Il prend sa source près de Souk-Ahras, dans le Constantinois algérien, puis coule vers l'est avant de se jeter dans la mer Méditerranée dans le golfe de Tunis. La Medjerda draine 23 600 km^2 dont 7500 km^2 en Algérie (Habaieb, 1992) (Fig. 1).

Avant de se jeter dans la mer, la Medjerda traverse trois types de plaines :

- en amont, la haute vallée (du côté de Ghardimaou) est constituée de bassins colmatés par des sédiments quaternaires. Le fleuve et ses affluents sont caractérisés par des méandres encaissés de 8 à 10 m. Ces régions appartiennent à des zones faiblement montagneuses. Les fortes crues ne débordent qu'exceptionnellement en dehors du lit du fleuve (Kotti et al., 2016) ;
- en aval de la retenue de Sidi Salem, de Testour à Jdaida, dite moyenne vallée de la Medjerda, le fleuve est encaissé dans la plaine. L'encaissement repasse à plus de 10 m en aval de Laroussia avant de s'atténuer rapidement vers Jdaida. Dans tout ce secteur la plaine est inondable surtout dans la partie centrale. En allant vers la mer les reliefs de ce bassin s'adoucissent ;
- dans la basse plaine, en aval de Jdaida, commence le delta. La pente de la Medjerda est encore marquée et son tracé a fréquemment été modifié, soit naturellement soit par des travaux (coupure de méandres, ouverture de nouvel émissaire...). Du point de vue géologique, la Medjerda est occupée par des alluvions quaternaires le plus souvent argilo-sableuses à argilo-sablo-limoneuses.

La pluviométrie moyenne annuelle dépasse 1000 mm au nord du bassin versant de la Medjerda et diminue progressivement vers le sud. Il s'agit d'une pluviométrie souvent concentrée dans le temps, irrégulière et inégalement répartie sur les différentes saisons (Kotti et al., 2016). La base des données hydrologique provient en grande partie de la Direction Générale

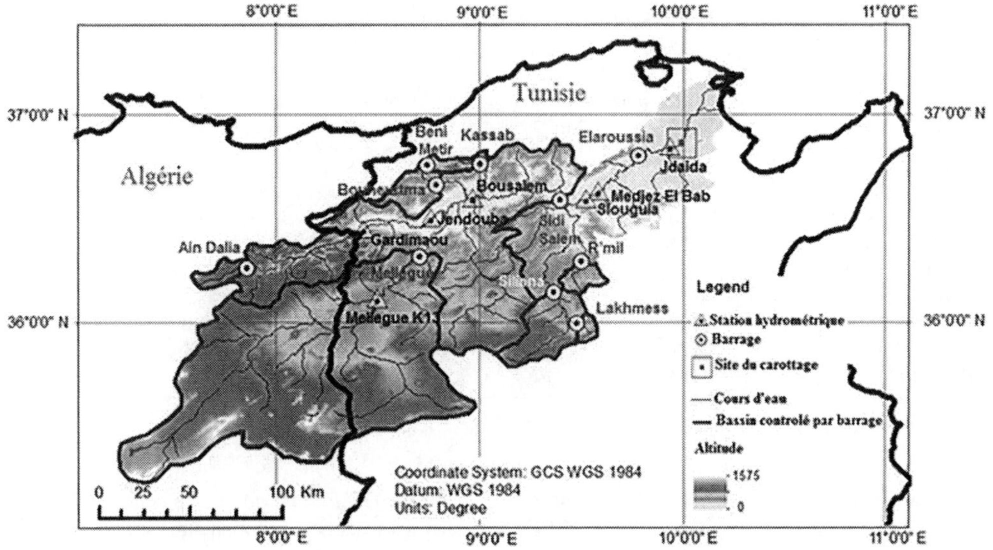

Figure 1. Localisation des grandes stations hydrométriques et des barrages sur le bassin versant de la Medjerda, et site du carottage.

des Ressources en Eau. Par ailleurs le laboratoire HydroSciences Montpellier (HSM) dispose d'une base de données pluviométrique importante au pas d'espace du demi-degré carré, qui propose des valeurs mensuelles de pluie par maille de 2500 km^2 sur la période 1940–1999, complétée par des données nationales (DGRE).

2.2 Régime hydrologique et stratégie de mobilisation des eaux de surface

Le régime hydrologique est typiquement méditerranéen. Les contrastes saisonniers sont très marqués puisque, pendant l'été, le débit d'étiage peut se réduire à moins d' 1 $m^3\,s^{-1}$ tandis que, pendant le reste de l'année, celui d'une crue de périodicité décennale atteint 1000–1200 $m^3\,s^{-1}$ (Rodier et al., 1981). Le régime hydrologique de ce fleuve très hétérogène a été régulé par de nombreux ouvrages construits sur son cours principal ou ses affluents. A partir de 1945, un programme d'équipement de la Medjerda a été entrepris par la Direction des Travaux Publics de Tunisie, destiné à mettre en valeur les ressources encore inutilisées de cette vallée. Ce programme a permis la construction du barrage régulateur de Mellegue et la construction du barrage de Beni Metir (Strugo, 1955). Depuis l'indépendance de la Tunisie, le programme de développement du bassin de la Medjerda a été poursuivi grâce à la réalisation de plusieurs ouvrages : Lakhmes (1966), Kasseb (1968), Bouhertma (1976), Siliana (1987), Sidi Salem (1981) et R'Mil (1999). En Algérie, un seul grand barrage a été identifié, Ain Dalia, mis en eau en 1988 et d'une capacité de 82 Mm^3. Avec la mise en eau de ces barrages, le régime hydrologique du fleuve Medjerda a été modifié (Fig. 1 pour la localisation et tableau 1).

2.3 Technique de prélèvements des sédiments, méthodes de mesure et géochronologie

Le site du carottage a été choisi après plusieurs missions de terrain (2013, 2014, 2015) qui avaient pour objectif d'identifier différents sites potentiellement intéressants pour la thématique. Le site d'étude est localisé dans un méandre en aval du dernier barrage -Sidi Salem- mais en amont de la zone estuarienne, proche de la station hydrologique Jdaida, la dernière avant la mer. Trois carottes ont été échantillonnées sur cette même terrasse alluviale, C1 de 90 cm, C3 de 120 cm et C5 de 300 cm de profondeur. L'approche paléohydrologique a été utilisée pour comparer les résultats sédimentologiques et géochronologiques de cette archive sédimentaire aux informations de la station hydrométrique de Jdaida située juste en amont du site d'étude choisi.

L'identification et la mise en évidence des différents événements de crues ont été réalisées grâce à l'inspection minutieuse de chaque dépôt (décompte des lamines, identification de paléosols, détection des surfaces d'érosion, étude de la variabilité granulométrique à partir d'un granulomètre laser Beckman Coulter LS 13320 au laboratoire Géosciences Montpellier.

Toute reconstitution paléo-environnementale, à partir de l'étude des archives sédimentaires, nécessite l'établissement d'une échelle chronologique la plus fine possible, afin de dater avec une bonne résolution les événements hydrologiques passés. Les isotopes principalement utilisés pour dater les événements du siècle dernier sont le ^{210}Pb et le ^{137}Cs (Dezileau et al., 2014a). Ils permettent une datation relativement précise, grâce à leurs propriétés chimiques (réactivité vis-à-vis des particules) et physiques (périodes de désintégra-

Tableau 1. Caractéristiques des aménagements construits sur le bassin versant de la Medjerda (DGBGTH, 2016).

BARRAGES/ Mise en eau	Superficie du bassin (km^2)	Capacité totale initiale (Mm^3)	Envasement des retenues Mesurée (Mm^3)	Taux d'envasement des retenues (%)	Taux d'envasement annuel des retenues (%)
MELLEGUE (1954)	10 300	182	122	67	1,36
BENI METIR (1954)	103	62	1	*	*
KASSEB (1966)	170	81	5	6,08	0,19
SIDI SALEM (1981)	17 885	814	171	21	0,86
BOU HEURTMA (1976)	457	117	5	4,65	0,16
SILIANA (1987)	972	70	37	52,2	3,25
LAKHMESS (1968)	127	8,22	1	12,2	0,37
RMIL (1999)	221	4	*	*	*
AIN DALIA (1988)	193	82	(64)	(22)	0,91 (ANBT)

* Données manquantes.

tion, rayonnement gamma). Le ^{137}Cs et le ^{210}Pb sont mesurés par spectrométrie gamma (détecteur CANBERRA BEGe 3825) au laboratoire Géosciences Montpellier (France) et au CNSTN à Sidi Thabet (Tunisie).

- Datation par le ^{210}Pbex (T1/2 = 22,3 ans)

Le Plomb 210 est un isotope radioactif naturel issu de la chaîne de désintégration de l'Uranium 238. L'Uranium 238, contenu naturellement dans la croûte terrestre, subit la série de désintégration suivante : Uranium 238 – Uranium 234 – Thorium 230 – Radium 226 – Radon 222. Le 222Rn (T1/2 = 3,8 jours) ainsi produit, se trouve sous une forme gazeuse. Lorsqu'il est produit à proximité de la surface, le gaz s'échappe dans l'atmosphère où il se transforme rapidement en Polonium 218 qui se désintègre en quelques minutes en ^{210}Pb. Ce ^{210}Pb, dit atmosphérique (ou en excès) retombe sous forme de particules humides ou sèches et s'accumule dans les sédiments. Son activité au sein de la colonne sédimentaire diminue avec le temps selon une loi de décroissance radioactive. Le principe de la datation au ^{210}Pb s'appuie sur l'hypothèse selon laquelle le flux atmosphérique en ^{210}Pb est constant, ce qui implique que l'activité de l'isotope radioactif à la surface des sédiments reste toujours identique. Cette hypothèse permet en mesurant l'activité du ^{210}Pb et en appliquant la loi de décroissance radioactive de déterminer l'âge de nos sédiments. Au bout de cinq fois sa période, c'est à dire 110 ans, il a complètement disparu dans l'environnement (Dezileau et al., 2014b). La mesure du ^{210}Pb sur le détecteur gamma CANBERRA se fait en intégrant le pic à 46,5 keV. Le ^{210}Pbex est estimé en soustrayant le pic du ^{210}Pb au pic du ^{226}Ra à 186,2 keV. L'erreur n'excède pas 6 %.

- Datation par le ^{137}Cs (T1/2 = 30,17 ans)

Le Césium 137 est un isotope radioactif artificiel issu de réactions de fission nucléaire. Il est relâché dans l'atmosphère lors de l'utilisation de bombes atomiques et d'accidents dans des centrales nucléaires. Avec une période de demi-vie (30,17 ans) proche de celle du ^{210}Pb, il permet de dater les séquences sédimentaires sur une même échelle temporelle. Les profils d'activité du Césium 137 en fonction de la profondeur dans une coupe sédimentaire présentent des pics dont le plus important est généralement attribué à l'année 1963, date du maximum de rejets atmosphériques par les essais nucléaires. La présence de Césium 137 dans les dépôts sédimentaires indiquera qu'ils ont été déposés postérieurement aux années 1950. La mesure du ^{137}Cs sur le détecteur gamma CANBERRA se fait en intégrant le pic à 661 keV. L'erreur n'excède pas 6 %.

3 Résultats

3.1 Sédimentation des réservoirs

Le taux d'envasement de chaque barrage construit sur le bassin versant de la Medjerda met en évidence l'importance de l'alluvionnement durant la durée de l'exploitation des retenues des barrages, dont Mellegue, Siliana et Sidi Salem qui sont sujettes à un alluvionnement plus important (tableau 1). Il a été estimé que le volume initial de la retenue de Sidi Salem sera envasé jusqu'à 60 % 100 ans après la construction du barrage, et l'envasement devrait être total pour les retenues des deux barrages de Mellegue et Siliana dans 60 à 65 ans (Ben Mammou et Louati, 2007). Des soutirages efficaces permettent dans ce cas de réduire le taux d'alluvionnement et d'augmenter la durée de vie de ces barrages. D'après Abid (1980), 59 à 64 % des apports solides dans la retenue du barrage de Mellegue, durant la période 1954–1980, ont été soutirés grâce aux manœuvres de dévasement pendant les crues. La Fig. 2 montre le volume de vase évacué suite à des événements exceptionnels entre 1954 et 2003 au niveau du barrage Mellegue et entre 1982 et 2005 au niveau du barrage Sidi Salem. En 1969, de très importants volumes de sédiments évacués lors d'une crue se sont accumulés surtout dans les zones où la pente diminue, ainsi que dans les secteurs évasés. En mars 1973, l'oued Medjerda a connu une autre

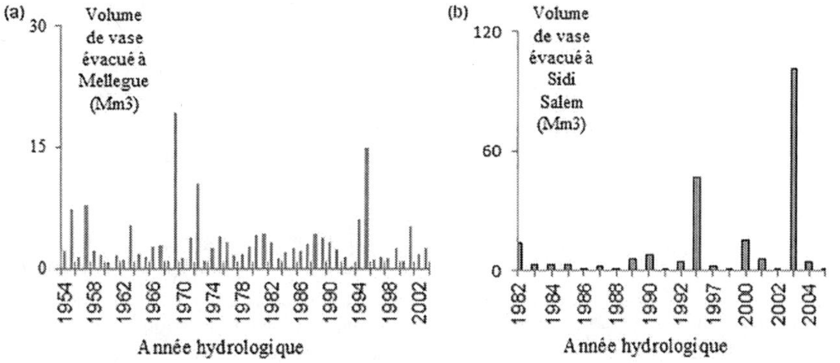

Figure 2. **(a)** Le volume de vase évacué au niveau du barrage Mellegue entre 1954 et 2003. **(b)** Le volume de vase évacué au niveau du barrage Sidi Salem entre 1982 et 2005.

crue plus intense provoquant de graves inondations avec une quantité exceptionnelle de matériaux transportée et ce malgré les gros travaux d'endiguement effectués à Jdaida (entre 1947 et 1957) (Ben Hassine and Rejeb, 2003).

Depuis la mise en eau du barrage Sidi Salem en 1981 jusqu'en 1990, le volume de vase évacué par ce barrage est quasiment nul. On remarque le faible pourcentage de vase évacuée dû à l'arrêt des opérations de soutirage. Quelques opérations de dévasement occasionnel ont été associées à des débits maximum exceptionnels : $595\,m^3\,s^{-1}$ en 1990, $671\,m^3\,s^{-1}$ en 1995, $977\,m^3\,s^{-1}$ en 2000 et $1020\,m^3\,s^{-1}$ 2003. Durant l'année 1995, les gestionnaires de Sidi Salem ont libéré un volume total dévasé de $24{,}27\,Mm^3$. Les particules solides se déposent en aval en fonction de la réduction de la vitesse du courant de turbidité, ainsi plusieurs facteurs peuvent influencer les épaisseurs des dépôts accumulés tels que la vitesse du courant, la topographie, la présence d'obstacles perturbant les dépôts, etc. Ces facteurs créent une variabilité latérale de dépôt importante, en aval de la retenue. Le lit de la Medjerda connaît de ce fait un engraissement continu depuis 1981, date de la mise en service du barrage Sidi Salem. De ce fait, et par manque de débits de crue important qui " chassent " les dépôts précédents, le débit occasionnant le débordement a diminué à moins de $200\,m^3\,s^{-1}$, valeur nettement inférieure au débit de débordement avant la construction du barrage qui était de $700\,m^3\,s^{-1}$.

La Fig. 2a illustre parfaitement l'accroissement des volumes solides évacués suite à la crue exceptionnelle de 2003. Quatre fortes crues très rapprochées sont survenues dans le bassin de la Medjerda au début de l'année 2003, après quatre années de sécheresse (Abdelhamid et Khalil, 2009). Elles ont cumulé un milliard de m^3 d'apports à Sidi Salem en un mois. La retenue a été assez vite remplie en début d'hiver. Ainsi, après complet encaissement de la première crue dans le creux disponible, et déversement de la seconde crue, la troisième a poussé le déversement sur la tour et réduit la tranche de laminage résiduelle (Daoud et al., 2009).

3.2 Modélisation et reconstitution des transports solides

L'objectif principal de cette section est de développer un modèle simple qui montre l'atténuation du signal des apports solides à Jdaida avec les données disponibles sans tenir compte des paramètres physiques du bassin. La méthodologie s'intéresse d'abord au premier barrage Mellegue construit puisque, on dispose à cette station de 1954 à 2005 d'un historique des apports solides annuels au barrage MES X_{Mellegue} et des volumes annuels de vase évacués MES lachées Y_{Mellegue}. A la construction de chaque barrage correspond un stockage de MES que l'on peut dans un premier temps estimer par un pourcentage Y_n d'une quantité de MES X_n. Le coefficient annuel de réduction de transport solide pour le barrage de Mellegue qui a été calculé est de l'ordre 51 %. Le fonctionnement sera étudié après pour tout le reste du bassin versant Medjerda, grâce à des valeurs calculées pondérées par la surface de chaque barrage en appliquant le même coefficient de réduction à Mellegue pour les autres barrages, ce qui constitue bien sur une hypothèse réductrice, mais basée sur la seule observation disponible.

Avant la mise en eau du barrage Sidi Salem en 1981, les barrages Mellegue, Beni Metir, Bouherma, Kassab et Lakhemes (tableau 2) déversaient vers la mer sans autre obstacle. Depuis 1981, les quantités de MES qui ressortent de ces barrages arrivent au barrage d'ordre 2 Sidi Salem, situé juste après la station Bousalem, sur le cours de la Medjerda (tableau 2). Depuis la mise en eau du barrage Siliana en 1987, les quantités qui ressortent du barrage Lakhemes arrivent au barrage Siliana d'ordre 2. Cette étude permet de classer l'ordre de grandeur des flux des matières solides jugés assez proche de la réalité. La quantité de MES qui arrive à Jdaida est calculée en additionnant les quantités de MES lâchées par chaque barrage avec les quantités de MES produites dans le bassin intermédiaire, en appliquant à chaque fois la réduction de MES déjà calculée à Mellegue. Les résultats de cette approch, illustrés par la Fig. 3 et le tableau 2, montrent la réduction de la quantité de MES calculé à la sta-

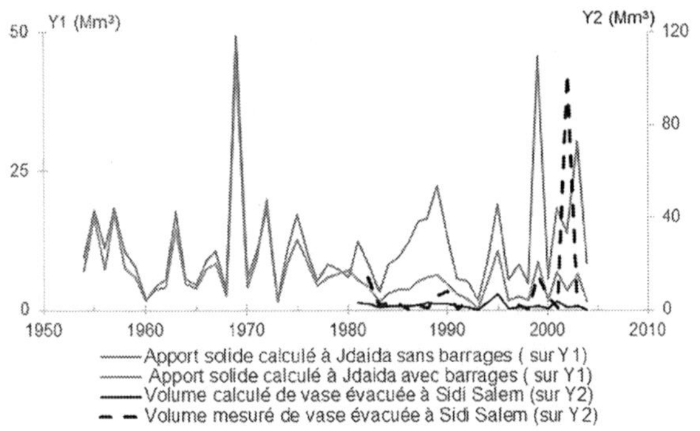

Figure 3. Evolution des apports solides calculés à Jdaida et des volumes calculés et mesurés de vase évacuée à Sidi Salem.

tion de Jdaida depuis la mise en eau du barrage Sidi Salem. Notons tout de même qu'il faut prendre les résultats obtenus avec beaucoup de précaution, cependant cela produit un signal d'atténuation des apports solides qui peut servir à interpréter les carottes prélevées dans la basse vallée. Nous avons eu recours à la modélisation à l'aide d'un modèle simplifié pour pallier au manque des données réelles dans le but de l'estimation du transport solide dans la Medjerda, mais il faudrait mettre en place un programme de mesures de débits solides au sortir des barrages pour valider ce modèle. Cette approche doit être approfondie en prenant en compte les informations qui pourront être obtenues auprès des gestionnaires de barrages, afin de mieux identifier la variabilité locale de la production de MES et des taux de sédimentation dans les barrages. D'autres modèles seront peut-être mieux appropriés avec beaucoup plus de précision concernant entre autre la production de sédiments. Il est recommandable de donner plus d'effort et d'investissements de la part des organismes concernés pour pouvoir donner une estimation exacte du transport solide.

3.3 Approche paléohydrologique pour évaluer l'impact anthropique

3.3.1 Reconstitution de la variabilité passée des transports sédimentaires du bassin versant de la Medjerda, Tunisie

L'utilisation de données granulométriques est indispensable à l'étude des archives sédimentaires. On se limite dans cette partie au résultat d'une seule carotte C5, la plus profonde et la plus riche en couches de sédiments différentes à interpréter. Les deux autres carottes C1 et C3 ne sont pas assez profondes pour atteindre le pic de Cesium, mais elles mettent en évidence une certaine variabilité latérale des dépôts, tout en confirmant la nature essentiellement argileuse des premiers 80 cm de dépôt.

La mise en place d'un log litho stratigraphique nous a permis d'identifier visuellement les différents dépôts de crues. L'archive sédimentaire montre d'importantes variations granulométriques en fonction de la profondeur, d'un matériel silteux-argileux à sableux (D50 compris entre 5 et 240 µm). Ces niveaux sableux sont d'un grain fin à moyen (de 63 à 449 µm) aux niveaux des profondeurs 125 cm, 150, 210, 250 et 290 cm, indiquant un régime hydraulique plus énergétique. Ces pics correspondent probablement à des événements exceptionnels (crues intenses).

L'ensemble des travaux d'observation, des études granulométriques et géochimiques de la carotte C5 permet d'identifier de manière plus fine le nombre de dépôts de crues. Les variations granulométriques et géochimiques sont le reflet des modifications de la dynamique sédimentaire. Le nombre de dépôts de crue est estimé à 18 et sont notés de FD1 à la base de l'archive à FD18 au sommet. Dans la séquence des dépôts de crue de la carotte C5, l'unité de dépôt FD1 présente une activité non négligeable de ^{137}Cs (Fig. 4) indiquant que tous les dépôts de crue de FD1 à FD18 sont tous postérieurs à 1950. D'autre part, la forte activité en cesium-137 mesurée au niveau de l'unité FD 7 (172 cm; 8,2 mBq g^{-1}), nous permet de faire la supposition que ce dépôt de crue est associé à la période de production maximale de Cesium-137 dans l'atmosphère, c'est à dire au milieu des années 1960 (autour de 1963, Fig. 4). Le ^{210}Pb est en excès dans l'unité FD1, ce qui indique que les dépôts de crue FD1- FD18 sont récents et tous postérieurs aux années 1900. D'autre part, nous avons trouvé sur le terrasse alluviale entre 120 et 140 cm, des déchets de pots de yaourth " Sami " fabriqué dans les années 1970.

Ces traceurs chronologiques, nous ont permis de mettre en place une chronologie approximative des dépôts de crue (FD1 à FD18) que nous comparons à l'historique des crues disponibles depuis 1930 à la station de Jdaida (Fig. 4). La comparaison de ces enregistrements permet d'attribuer de manière plus fine des âges aux dépôts de crues de la séquence FD. L'événement de 1969, le premier événement puissant après 1963 enregistré par les archives instrumentales (832 $m^3\,s^{-1}$), a probablement déposé une unité de texture différente en silt limite sables fins (FD8 : 44 µm). Un autre événement plus puissant encore, est enregistré en 1973 avec un débit estimé à 3500 $m^3\,s^{-1}$ à Medjez El Bab. Les mesures hydrologiques réalisées pendant la crue montrent qu'un débit de 850 $m^3\,s^{-1}$ est passé dans le lit du fleuve, mais que le reste a débordé déclenchant des inondations exceptionnelles (Claude et al., 1977). Si nous supposons qu'il existe une corrélation entre l'intensité d'un événement et la taille granulométrique qui caractérise son dépôt (Dezileau et al., 2014b), alors nous pourrions associer l'événement de 1973 à l'unité FD10 caractérisée par des sables fins (FD10 : 59 µm, 20 cm d'épaisseur). Cela est cohérent par rapport aux datations au ^{137}Cs et la découverte dans ces dépôts de pots de yaourt " Sami ". En nous fondant sur une relation possible entre la granulométrie des sédiments et l'ampleur d'un événement, FD17 (FD 7 : 100 µm) pourrait être associé à 2003, pour un débit 750 $m^3\,s^{-1}$ à Jdaida et FD 12 avec du maté-

Tableau 2. Rapport sur valeur moyenne de MES calculé au niveau de chaque barrage et des valeurs moyennes de MES calculées à Jdaida.

		valeur moyenne de MES calculé (m^3)	
Barrage	Ordre de grandeur des flux des des matières solides	stockée dans le barrage	lâchée par le barrage
Mellegue	Barrage d'ordre 1	2 136 732	3 269 028
Beni Metir	Barrage d'ordre 1	21 359	32 677
Lakhemes	Barrage d'ordre 1	28 770	42 291
Kassab	Barrage d'ordre 1	38 070	56 020
Bouhertma	Barrage d'ordre 1	131 856	144 417
Sidi Salem	Barrage d'ordre 2	2 169 357	1 901 540
Siliana	Barrage d'ordre 2	300 445	258 208
Passant à Jdaida	avec les barrages	6 718 661	
Passant à Jdaida	sans les barrages	11 503 176	

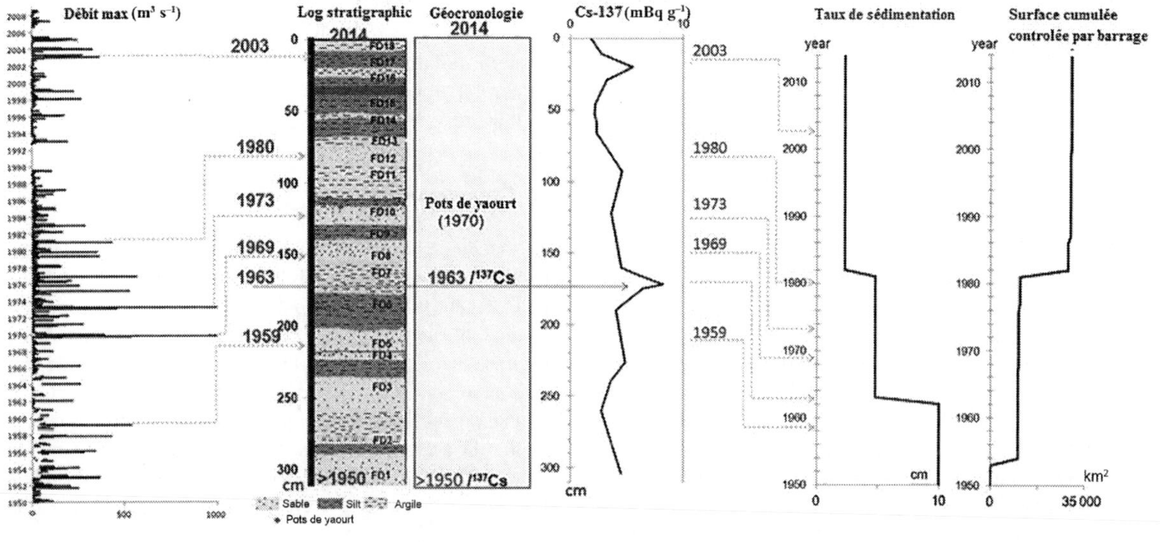

Figure 4. Reconstitution historique des apports sédimentaire par les traceurs chronologiques (^{137}Cs) et la comparaison des enregistrements de débit maximum annuel au niveau de la station hydrométrique de Jdaida.

riel fin (D50 of 26 µm) à 1980 (Fig. 4). D'autres unités sous-jacentes FD5 avec D50 de 212 µm, FD4 avec D50 de 212 µm et FD1 avec D50 de 224 µm ont été affectées à quatre événements de plus faible ampleur que celui de 1973 (538, 432 et 344 $m^3\,s^{-1}$) qui ont eu lieu en 1959, 1957 et 1956 (ou 1952) respectivement (Fig. 4). Les couches sédimentaires associées à ces crues sont caractérisées par des sables moyens (211 à 228 µm). Ceci est expliqué par le fait qu'avant 1969, les sections du cours d'eau étaient dégagées, la terrasse alluviale de la zone d'étude était en évolution. Cette approche nous a permis de tracer l'histoire sédimentaire à travers l'étude de la carotte de sédiments prélevés. Les résultats indiquent un dépôt de 75 cm entre 1981 et 2014, alors qu'il était de 230 cm entre 1956 (ou 1952) et 1981 (120 cm entre 1956/52 et 1963, puis 105 cm entre 1963 et 1981). L'histoire sédimentaire du fleuve Medjerda montre une réduction du flux sédimentaire à l'aval du barrage de Sidi Salem du fait des multiples aménagements.

3.3.2 Etude de l'évolution des apports : impact anthropique

La plaine de Medjerda est l'une des zones les plus peuplées de Tunisie, 288 834 habitants en 2004 (INS). Depuis l'indépendance en 1956, le pays a connu une croissance rapide de la population, et les zones urbaines au nord ont augmenté de 7,16 % entre 1950 et 2007 (Samaali, 2011). Compte tenu de l'augmentation des utilisations de l'eau dans les pôles économiques urbains au nord-est et au centre-est, l'état a suivi un plan de développement pour la mise en place d'un grand

Tableau 3. Evolution des sédiments transportés dans la basse vallée de Medjerda depuis 1950.

Année	1950–1963	1963–1981	1981–2014
Nombre de grands barrages	2	5	9
Surface cumulée contrôlée km^2	10 304	11 161	30 445
Taux de transport sédimentaire $cm\,an^{-1}$	10	4,75	2,34
Nature de sédiment	sable fin et sable grossier	silt et sable fin	silt ou argile

nombre de retenues, dont 9 grands barrages construits dans le bassin versant de la Medjerda, afin de répondre à l'intérêt général.

Les résultats d'analyse granulométrique de la carotte C5 montrent principalement une succession de petites couches de matériel fin (silt ou argile) depuis la mise en eau du barrage Sidi Salem en 1981 et un taux de sédimentation faible (2,3 $cm\,an^{-1}$), alors qu'il était beaucoup plus important, environ 4.75 $cm\,an^{-1}$ entre 1963 et 1981, et environ 10 $cm\,an^{-1}$ entre 1950 et 1963 (tableau 3).

Ceci illustre bien l'impact de la mise en eau des nombreux barrages sur le bassin, qui prive la basse vallée, l'estuaire et la zone littorale de la majeure partie des apports sédimentaires, en particulier grossiers.

4 Discussion

Généralement, les barrages limitent les apports de sédiments à la mer et participent au déficit général de sédiments sur les côtes. Ils capturent jusqu'à 30 % des flux de sédiments produits sur les massifs (Vorosmarty et al., 2003). Depuis la construction des barrages sur le bassin versant de la Medjerda, dont le plus important, celui de Sidi Salem, le lit de la Medjerda a connu d'importants changements morphologiques dans sa basse vallée, et en particulier un important engraissement du lit mineur. Cet engraissement est dû en grande partie à la réduction des débits de crue exceptionnels qui devraient " chasser " les dépôts anciens du lit mineur. Ceci a diminué la capacité de transit du fleuve ainsi que la vitesse d'écoulement entre la sortie du barrage et l'estuaire. Il en résulte l'apparition de débordements plus fréquents et la propagation des inondations pour des débits de plus en plus faibles. Par ailleurs, les crues ont également beaucoup moins d'impact mécanique sur la région côtière.

Les résultats de mesures de l'alluvionnement menées depuis une vingtaine d'année, montrent l'importance de la sédimentation dans les retenues des barrages de Tunisie. Des concentrations supérieures à 100 $g\,L^{-1}$ ont été enregistrées dans les eaux lors des crues de la Medjerda (Rodier et al., 1981). De ce fait, les retenues créées par les différents ouvrages hydrauliques sont toutes confrontées à plus ou moins long terme à l'alluvionnement. Les mesures de transport solide effectuées en Tunisie et les études ponctuelles détaillées restent insuffisantes et n'ont pas été menées assez longtemps pour permettre des études exhaustives sur le sujet. La modélisation du transport sédimentaire proposée dans cette étude résulte donc d'une approche simplifiée destinée à montrer l'atténuation du signal des apports sédimentaires, calculé à partir de données fragmentaires de transport solide et de taux d'envasement disponibles sur quelques sites d'étude sur le bassin. Cette première approche indique une réduction croissante des transports sédimentaires avec l'augmentation du nombre de barrages, qui peut servir également à supporter l'interprétation des archives sédimentaires prélevées dans les carottes de la basse vallée.

5 Conclusion

La Medjerda est devenu un fleuve presque totalement contrôlé, et les matières solides qu'il apporte ont été réduites de façon considérable. Ce qui indique aussi que les crues ont beaucoup moins d'impact mécanique sur la région côtière. La mise en évidence de différents paléo-événements de crue qui ont été datés par des méthodes complémentaires (^{137}Cs, $^{210}Pbex$, géochimie, historiques des crues) montre que la terrasse FD a enregistré l'ensemble des événements de crue entre 1952 (ou 1956) et 2014. Une succession visible de couches sédimentaires correspondant aux dépôts des différentes crues qui se sont succédées sur le site d'étude a été déterminée. Elle permet de reconstituer l'histoire des apports sédimentaires du fleuve Medjerda jusqu'à la mer sur une période de 60 ans, avant et après la construction de nombreux barrages. Cette approche a permis d'évaluer le taux de réduction et la variabilité interannuelle des transports des sédiments. Toutefois, il faudrait une étude multi-site pour valider cette approche.

D'après la carotte C5, le taux de sédimentation était d'environ 10 $cm\,an^{-1}$ entre 1952/56 et 1963, principalement constitue de sables, puis il est passé à environ 4.75 $cm\,an^{-1}$ entre 1963 et 1982, avec un équilibre plus net entre sables, silts et argiles. Après la mise en eau du barrage Sidi Salem en 1981 sur le cours d'eau principal, les apports consistaient principalement en matériel fin, avec un taux de sédimentation faible (2,3 $cm\,an^{-1}$), Il s'agit d'une diminution significative des apports sédimentaires à la mer, et d'une modification du type de sédiments, essentiellement fins avec une disparition du matériel grossier, dû aux barrages, et en particulier celui de Sidi Salem sur le fleuve Medjerda. Actuellement la surface contrôlée par les barrages existants sur le bassin de la Medjerda est de 130 % de la surface du bassin de la Medjerda, ce qui veut dire que l'eau de ruissellement est stockée 1,3 fois sur son parcours vers l'estuaire, avec les conséquences enregistrées dans la carotte sur la dynamique sédimentaire pré-estuarienne. D'autres études (Benmoussa et al., 2017), illustrent l'impact de ce déficit d'apports sédimentaires dans la morphodynamique côtière du golfe de Tunis.

Disponibilité des données. Des données sont disponibles sur demande.

Intérêts concurrents. Les auteurs déclarent qu'ils n'ont acun conflit d'intérêts.

Déclaration de la numéro special. This article is part of the special issue "Water quality and sediment transport issues in surface water". It is a result of the IAHS Scientific Assembly 2017, Port Elizabeth, South Africa, 10–14 July 2017.

Remerciements. Nous remercions l'Institut National Agronomique de Tunisie et son Directeur Monsieur Mahmoud Elyes Hamza, pour le soutien logistique et financier, et pour l'octroi d'une bourse de stage à l'Université de Montpellier. Le programme Campus France est particulièrement remercié pour l'attribution d'une bourse Eiffel de séjour long en France, ainsi que l'IRD pour une bourse de séjour scientifique long en France. Nous remercions aussi le Centre National des Sciences et Technologies Nucléaires pour la disponibilité du matériel de carottage.

Edited by : Christophe Cudennec
Reviewed by : Didier Orange and one anonymous referee

Références

Abdelhamid, D. and Khalil, J. : Couplage d'un évacuateur vanné avec une tranche de laminage, cas du barrage de Sidi Salem en Tunisie, Colloque CFBR-SHF : "Dimensionnement et fonctionnement des évacuateurs de crues", 20–21 janvier, Lyon, 2009.

Abid, A. : Apports solides et soutirages aux barrages Nebeur sur l'oued Mellegue (1954–1980), Séminaire International d'Experts sur le Dévasement des retenues, Tunis, 1–4 juillet, 13 pp., 1980.

Baker, V. R., Webb, R. H., and House, P. K. : The scientific and societal value of paleoflood hydrology, in : Ancient Floods, Modern Hazards : Principles and Applications of Paleoflood Hydrology, edited by : House, P. K., Webb, R. H., Baker, V. R., Levish, D. R., Water Science and Application Series, 5, 127–146, 2002.

Bates, B. C., Kundzewicz, Z. W., Wu, S., and Palutikof, J. P. : Le changement limatique et l'eau, document technique publié par le Groupe d'experts intergouvernemental sur l'évolution du climat, Secrétariat du GIEC, Genève, 236 pp., 2008.

Ben Hassine, H. and Rejeb, M. M. : Crues et inondations dans la moyenne Mejerda, Rapport de synthèse DGRE, 2003.

Ben Mammou, A. and Louati, M. H. : Evolution temporelle de l'envasement des retenues de barrages Tunisie. Revue des sciences de l'eau, Journal of water science, 20, 201–210, 2007.

Ben Mansoura, A., Garchi, S., and Daly, H. : Analyzing forest users' destructive behavior in Northern Tunisia, Land Use Policy, 18, 153–163, 2001.

Benmoussa, T., Amrouni, O., Dezileau, L., Mahe, G., and Abdeljaouad, S. : Impact of the Medjerda sedimentary fluxes on the morphodynamic equilibrium of the northern coast of the Gulf of Tunis (Medjerda-Raoued coast), Tunisia, PIAHS 377, IAHS meeting, Port Elizabeth, South Africa, 2017.

Claude, J., Francillon, G., and Loyer, J. Y. : Les alluvions déposées par l'Oued Medjerda lors des crues exceptionnelles de Mars 1973, Cah. ORSTOM, sér. Hydrol., vol. XIV, no. II, 1977.

Daoud, A. and Jemmali, K. : Couplage d'une vacuateur vanne avec une tranche de laminage, cas du barrage Sidi Salem en Tunisie Combining gated spillway and flood damping, Sidi Salem dam in Tunisia. Colloque CFBR-SHF : "Dimensionnement et fonctionnement des évacuateurs de crues", 20–21 janvier 2009, Lyon – Daoud Jemmali Goguel Leclerc, Couplage d'un évacuateur vanné avec une tranche de laminage, cas du barrage de Sidi Salem en Tunisie, 2009.

Dezileau, L., Terrier, B., Berger, J. F., Blanchemanche, P., Freydier, R., Bremond, L., Latapie, A., Paquier, A., Lang, M., and Delgado, J. L. : A multidating approach applied to historical slackwater flood deposits of the Gardon River, SE France, Geomorphology, 214, 56–68, https://doi.org/10.1016/j.geomorph.2014.03.017, 2014a.

Dezileau, L., Terrier, B., Berger, J. F., Blanchemanche, P., Latapie, Freydier, R., Paquier, A., Lang, M., and Delgado, J. L. : Paleohydrological reconstruction of extreme floods of the Gardon River, SE France, Houille Blanche, International Water Journal, 4, 44–52, 2014b.

Ghorbel, A. and Claude, J. : Mesure de l'envasement dans les retenues de sept barrages en Tunisie : estimation des transports solides, DRES-ORSTOM, No. : 049 34, Tunis, 1982.

Habaieb, H. : Comparaison numérique des modèles de prévision des crues, application à des bassins versants belge, français et tunisien. Doctorat d'Etat en Sciences Agronomiques de l'Université de Gand, Belgique, Mai 1992.

Huntington, T. : Evidence for intensification of the global water cycle : Review and synthesis, J. Hydrol., 319, 83–95, 2006.

Kotti, F., Mahe, G., Habaieb, H., Dieulin, C., and Calvez, R. : Etude des pluies et des débits sur le bassin versant de la Medjerda, Tunisie, Bulletin de l'Institut Scientifique, 38, 19–28, 2016.

Labat, D., Godderis, Y., Probst, J., and Guyot, J. : Evidence for global runoff increase related to climate warming, Adv. Water Resour., 27, 631–642, 2004.

Rodier, J., Colombani, J., and Kallel, R. : Monographie de la Medjerda, ORSTOM, DGRE, 1981.

Samaali, H. : Etude de l'évolution de l'occupation et de l'utilisation du sol dans le delta de Medjerda par télédétection et systèmes et systèmes d'informations géographiques, Thèse en vue de l'obtention du Doctorat en Géographie. Université de Tunis Faculté des Sciences Humaines et Sociales Ecole Doctorale : Structures, Systèmes et Modèles en Lettres et Sciences Humaines, 2011.

Selby, K. A. and Smith, D. E. : Late Devensian and Holocene sea-level changes on the Isle of Skye, Scotland, UK, J. Quaternary Sci., 22, 119–139, 2007.

Solomon, S. : Climate change 2007 : the physical science basis, Cambridge University Press Cambridge, 2007.

Strugo, N. : La mise en valeur de la valeur de la vallée de la Medjerda, BEST, 102, 49–95, 1955.

Vorosmarty, J. V., Meybeck, M., Fekete, B., Sharma, K., Green, P., and Syvitski, J. P. M. : Anthropogenic sediment retention : major global impact from registered river impoundments, Global Planet. Change, 39, 169–190, 2003.

Vilanova, I., Prieto, A. R., and Espinosa, M. : Palaeoenvironmental evolution and sea level fluctuations along the southeastern Pampa grasslands coast of Argentina during the Holocene, J. Quaternary Sci., 21, 227–242, 2006.

Zahar, Y., Ghorbel, A., and Albergel, J. : Impacts of large dams on downstream flow conditions of rivers : Aggradation and reduction of the Medjerda channel capacity downstream of the Sidi Salem dam (Tunisia), J. Hydrol., 351, 318–330, 2008.

Proc. IAHS, 377, 77–81, 2018
https://doi.org/10.5194/piahs-377-77-2018
© Author(s) 2018. This work is distributed under
the Creative Commons Attribution 4.0 License.

The sedimentological changes caused by human impact at the artificial channel of Medjerda-River (Coastal zone of Medjerda, Tunisia)

Thouraya Benmoussa[1], Oula Amrouni[2], Laurent Dezileau[3], Gil Mahé[4], and Saâdi Abdeljaouad[1]

[1]University of Tunis El-Manar, Faculty of Science,
Laboratory of Mineral Resource and Environment, Tunis, Tunisia
[2]National Institute of Marine Sciences and Technologies,
Laboratory of Marine Environment, Tunis, Tunisia
[3]University of Montpellier II, HDR-Géosciences, Montpellier, France
[4]UMR HydroSciences Montpellier/IRD, Montpellier, France

Correspondence: Thouraya Benmoussa (benmoussa_thouraya@yahoo.fr)

Received: 13 June 2017 – Revised: 6 December 2017 – Accepted: 22 December 2017 – Published: 16 April 2018

Abstract. Recent sedimentary and morphological changes at the new mouth of Medjerda-River (Gulf of Tunis) are investigated using a multiproxy approach of sediment cores complited by ^{210}Pbex and ^{137}Cs method dating. The subject of the study is to focus on surveying the sedimentary evolution of Medjerda-Raoued Delta caused by the human intervention in the management of the main tributaries of the Medjerda-River (artificial channel of Henchir Tobias). Sediment cores (CEM-1 and CEM-3) were subjected to both multiproxy approaches (Grain size, geochemical analysis and dating radiometric ^{210}Pbex and ^{137}Cs). The sedimentological analysis of the new deltaic deposits shows a progradation sequence with the silt and clay deposits on the historic sandy substratum. The mean grain size evolution on the old beach profile shows a decreasing trend from backshore (CEM-3) to nearshore (CEM-1). The geochemical results show varying concentrations of chemical elements such as Fe, K, Rb, Nb, Cr, Ti, Ba, Ca, Sr, Zr, V, and potentially toxic metal trace elements such as Pb, Zn and the As. The Principal component Analysis (PCA) applied in the geochemical elements evolution confirms the marine origin of the sand deposits in the basic layers of the two cores. The chronological method (^{210}Pbex and ^{137}Cs) affirms that the first fluvial deposits were set up only after 1950. The sedimentological and geochemical result confirm the actual unless of coarser fluvial supplies under the human activities leading the negative coastal sediment balance and the shoreline retreat as well.

1 Introduction

The sediment budget of the coastal area is subject to several factors. The natural one is related to the sand supply sources and the eolian/hydrodynamics regime. The human factor represents an important element on the equilibrium of the coastal budget. Considering the coastline in its three dimensions, some authors have concentrated on the geomorphological balances of sedimentary flows, cell per cell (Suanez and Sabatier, 1999; Amrouni et al., 2014). The scientific community has also raised the question of the natural and human factors impacton the coastal sandy beach supplies (Flemming and Hansom, 2011).

The study of coastal environments linked to different global rivers is all the more important as they can be considered as markers of the evolution of sedimentary contributions in marines environments in relation to climatic as well as human variations (Milliman and Meade, 1983; Syvitski and Kettner, 2011; Kotti et al., 2015). The evolution of the nearshore sedimentary distributionis also related to the variability of hydrodynamic forcing and sediment availability in coastal areas (Perillo, 1995). The study area constitutes the coastal alluvial plain of the Medjerda River, the western bay

Published by Copernicus Publications on behalf of the International Association of Hydrological Sciences.

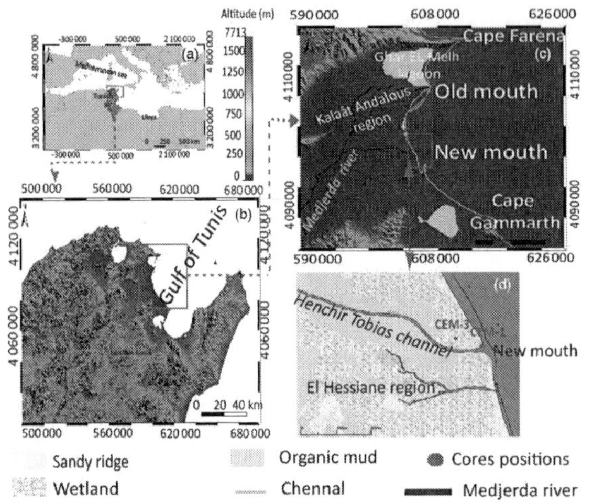

Figure 1. Map of study area location and cores sampling, deltaic coastal plain of the new mouth of Medjerda river, gulf of Tunis, NE of Tunisia. **(a)** Position of Tunisia in the Mediterranean Sea; **(b)** Northern-East of Tunisia; **(c)** topographic map of the gulf of Tunis, **(d)** the Sediment cores positions at the new mouth of Medjerda River.

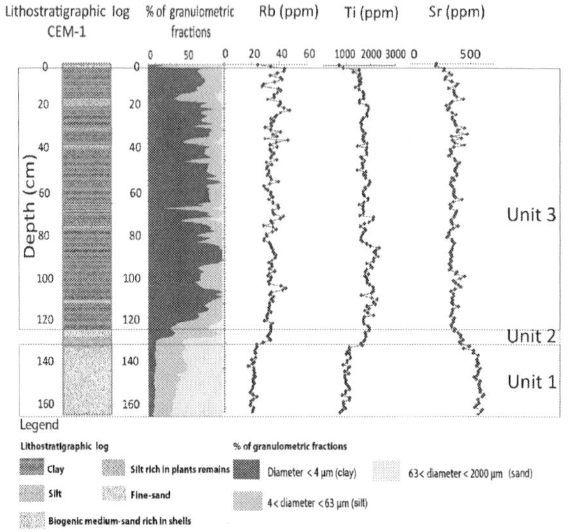

Figure 2. Grain size and geochemical elements evolution of deltaic plain of new mouth of Medjerda River (CEM-1).

of the gulf of Tunis. It is an estuarine coast, which receives the majority inputs feeding from the terrestrial hinterland (Oueslati et al., 2004). The anthropogenic activity mainly influencing the evolution of the northern coast sediment distribution of the Gulf (Saïdi, 2013; Saïdi et al., 2014).

The subject of this study is to describe the spatio-temporal impact of the management of the river flume on the sedimentological characteristics at coastal deltaic artificial channel of Medjerda River. We use a combined approach of sediment cores analysis and the ^{210}Pb and ^{137}Cs method dating to follow the creation and development of the new delta of Medjerda River (anthropogenically modified since 1972) (Soussi, 1988).

2 Study area

The study area, the Medjerda coast is located in the western bay of the gulf of Tunis between 37°10′ N–10°16′ E (Cape Farina) and 37°55′ N–10°18′ E (Cape Gammarth). The coastis characterized by three main sedimentary systems with three morphological entities: Lagoons of Kalaât El Andalous, Delta of Medjerda and the foredunes of Raoued in the El Hessiane areas (Fig. 1).

3 Material and methods

The methodology was based on sedimentological investigation completed by geochemical analysis on the new mouth of Medjerda River. Two short sedimentary cores CEM-1 and CEM-3 were collected from the Medjerda delta in the northeast of the Gulf of Tunis (Fig. 1c and d). The sampling protocol adopted consists of collecting 1.68 m and 45 cm, respectively for the two cores of deep sediment, in order to reconstitute a fine sedimentary evolution on the new mouth of Medjerda River evolution. A high resolution multi-proxy study was undertaken on the cores in the UM, HDR-laboratory, France. The granulometric analysis was carried out using a Beckman Coulter © LS 13 320. The geochemical analysis was conducted on the sedimentary material using the X-ray Fluorescence.The Principal Component Analysis (PCA) was carried out under the XLSAT-2016 software. Based on ^{210}Pb and ^{137}Cs by Gamma spectrometry in order to establish a chronology on the sedimentary evolution of the new delta of Medjerda-River during the last century.

4 Results and discussion

4.1 Grain size and geochemical evolution of sediment cores

The high-resolution (1 cm) grain size analysis of the cores shows a spatial variability. The CEM-1 core shows three distinctive sedimentary units (Fig. 2). The layer of fine sand (D50 = 98 µm), is surmounted by mixture of fine sand and silt laminates (Unit 2). But, the upper layer (Unit 3) is composed by clay (Unit 3). The CEM-3 core reveals the same sedimentary units (Unit 1, Unit 2 and Unit 3) (Fig. 3).

The geochemical results show varying concentrations of chemical elements such as Fe, K, Rb, Nb, Cr, Ti, Ca, Sr, Zr, V, and potentially toxic metal trace elements such as Pb, Zn in the sedimentary cores. The evolution of the contents of

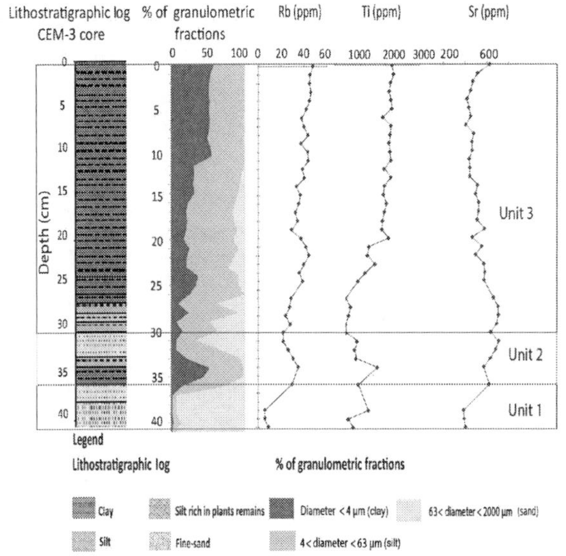

Figure 3. Grain size and geochemical elements evolution of alluvial plain of new mouth of Medjerda River.

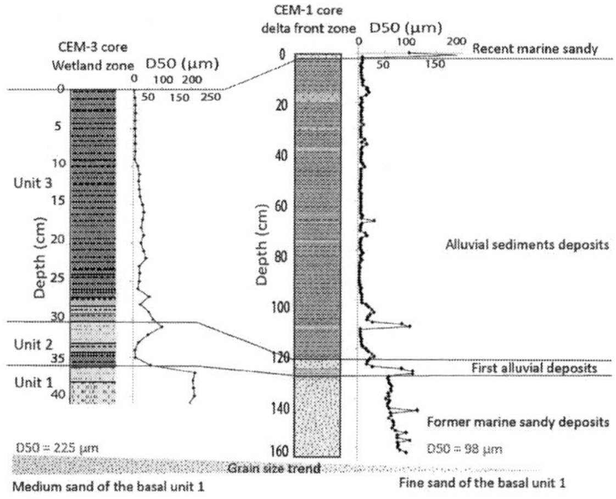

Figure 4. The mean grain size evolution of the old sandy deposits from the wetland (CEM-3) to the delta front (CEM-1) of the Medjerda River.

chemical elements in the sedimentary column reveals a good correlation with the grain size analysis. The marine tracers (Ca, Sr) show a significant decreasing concentration in the clay layers compared to their rate in the sandy one. While to the rate in terrigenous elements is higher in the fine sediments (clay and silt) (Figs. 2 and 3).

The Unit 1 of CEM-1 formed by fine-sand (D50 = 98 µm). But the Unit 1 of CEM-3 composed by medium-sand (D50 = 228 µm). The mean grain size (D50) evolution on the historical beach profile (before the alluvial deposits processes) shows a decreasing trend from the backshore (CEM-3) to the nearshore (CEM-1) (Fig. 4).

4.2 Correlation of geochemical elements

The Principal Component Analysis (PCA) of the geochemical elements shows the presence of 3 poles (Fig. 5): a pole of terrigenous elements (Fe, K, V, Rb, Cr, Ti), a pole of marine elements (Ca, Sr) and a pole of the pollutant minerals (Zn, Pb and As). A good correlation established between the elements present in the same pole. An anti-correlation observed between the first (fluvial elements) and second pole (Marine elements).

The grain size and geochemical analysis of the sediment cores shows that the finest detrital deposits (silt and clay) manifested the progradation of a new delta at the new mouth of the artificial channel. The grain size correlation established from the backshore to the nearshore shows that the historic beach was behind the present shoreline and the nearshore area was located on the current beach. It is a delta progradation on the old sandy beach. The retained fluvial discharge by the dam construction decreases the coarser fraction delta front toward silt to clay deposit.

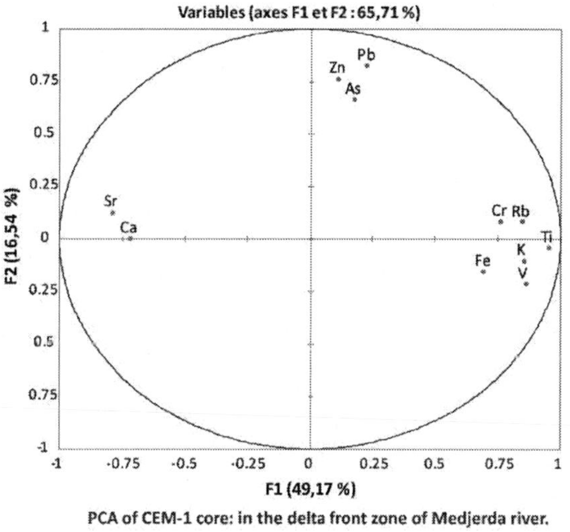

Figure 5. Principal Component Analysis of geochemical elements of the CEM-1 core sediments.

4.3 Historical evolution of the new delta of the Medjerda

The $^{210}Pb/^{137}Cs$ method dating shows that a new delta has been developed at the new mouth of the Medjerda since 1950. The Cesium 137 is an artificial radioactive element produced by the atmospheric nuclear tests. It is foundin the atmospheric environment from the years 50. The contents relatively important of cesium-137 measured in the units CEM-1-Niveau 80 cm (8.52 mBq g^{-1}) can be associated with

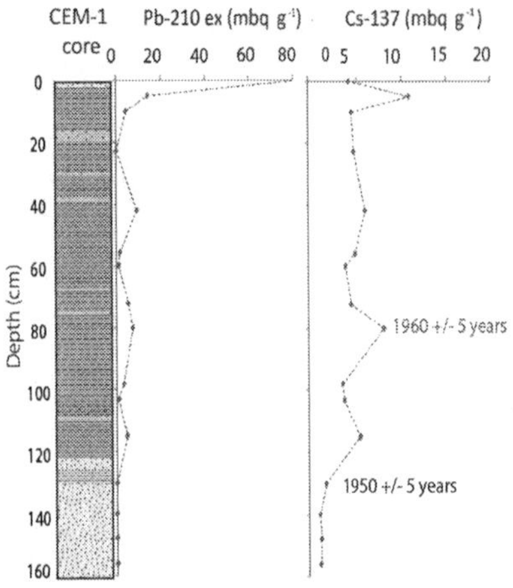

Figure 6. Geochronologyof sedimentary deposits at the new mouth of the Medjerda River.

the maximal production of cesium-137 atmospheric in the middle of 1960s (Fig. 6). However, The variability of content ^{137}Cs, along the CEM-1 core is the result of the mobilisation of this element especially in the marine and coastal environments, which does not allow to fix easily the real peak of maximum of nuclear broadcast (emission, issue) of 1963 (Dezileau et al., 2014). As a result, the relatively high cesium-137 levels measured in the CEM-1-Level 80 cm units ($8.52\,\mathrm{mBq\,g^{-1}}$) can be associated with the maximum production of atmospheric cesium-137 in the mid-1960s. The first peak which appeared at the base of the CEM-1 core at -115 cm results from the appearance of ^{137}Cs contained in aged layers of the appearance of this element in the atmosphere (before 1950).

5 Conclusions

The result shows a relatively rapid deltaic plain progradation ($1.53\,\mathrm{cm\,yr^{-1}}$) with a fluvial dominance marked by silt and clay deposits rich in terrigenous chemical elements from the year 1950. This sedimentary evolution of the new delta is manifested by the filling of the old mouth since 1973 flood. The winnowing of finest fraction (clay and silt) provided during the flood event on the delta front, by the active wave current ledto the preferential spatial grain size distribution toward the northern behavior coast.

The dominance of clay deposits more than coarser fraction result to the impact of the dam construction of several dams. Only the finer grain size fractions can feed the coast. The actual sediment balance of the gulf of Tunis coast shows a shoreline retreat under the coarser sediment deficit blocked by upstream dam construction.

Data availability. No data sets were used in this article.

Competing interests. The authors declare that they have no conflict of interest.

Special issue statement. This article is part of the special issue "Water quality and sediment transport issues in surface water". It is a result of the IAHS Scientific Assembly 2017, Port Elizabeth, South Africa, 10–14 July 2017.

Acknowledgements. The present study was investigated under the bilateral Tunisan-French project Utique-CMCU-RYSCMED16G1005 (2016-2018).

Edited by: Hafzullah Aksoy
Reviewed by: two anonymous referees

References

Amrouni, O., Hermassi, T., AbdelJaouad, S., and Messaoudi, S.: Contribution of Grain-size Trend to Sediment of a Microtidal Beach. Case of the Gulf of Tunis Bay (Cape Ferina-Cape Gammarth, Tunisia), Research Journal of Environmental Sciences, 8, 161–177, https://doi.org/10.3923/rjes.2014.161.177, 2014.

Dezileau, L., Terrier, B., Berger, J. F., Blanchemanche, P., Latapie, A., Freydier, R., Paquier, A., Lang, M., and Delgado, J. L.: Reconstitution des crues extrêmes du Gardon à partir d'une analyse paléohydrologique, Houille Blanche, EDP Sciences, 4, 44–52, https://doi.org/10.1051/lhb/2014037, 2014.

Flemming, B. W. and Hansom, J. D.: Estuarine and Coastal Geology and Geomorphology – A Synthesis, in: Treatise on Estuarine and Coastal Science, edited by: Wolanski, E. and McLusky, D. S., 3, 1–5, Waltham: Academic Press, 2011.

Kotti, F., Dezileau, L., Mahe, G., Habaieb, H., and Dieulin, C.: Reconstitution historique des apports sédimentaires par l'oued Medjerda, Tunisie, International Conference on African Large River Basins Hydrology, 26–30 October 2015, Hammamet, Tunisia, p. 41, 2015.

Milliman, J. D. and Meade, R. H.: World-Wide Delivery of River Sediment to the Oceans, J. Geol., 91, 1–21, doi.org/10.1086/628741, 1983.

Oueslati, A.: Littoral et aménagement en Tunisie, ORBIS presses, p. 534, 2004.

Perillo, G. M. E.: Geomorphology and sedimentology of estuaries, Developments in sedimentology, chap. 1, An Introduction, Vol. 53, 1–16. 1995.

Saïdi, H.: Etude sédimentologique et morphodynamique des côtes sableuses du golfe de Tunis (Tunisie nord-orientale), Thèse de doctorat, Faculté des Sciences de Tunis, 217 pp., 2013.

Saïdi, H., Souissi, R., Louati, M., and Zargouni, F.: Morphologic changes and sedimentary budgets along a Mediterranean coastline with a sand spit: case of the littoral fringe Sidi Ali El Mekki–Gammarth (NE Tunisia), Rend. Fis. Acc. Lincei, 25, 393–401, https://doi.org/10.1007/s12210-014-0314-0, 2014.

Soussi, N.: Les mécanismes de la sédimentation dans le golfe de Tunis (Tunisie), condensés des travaux présentés lors du *XXXIème congrès-Assemblée plénière*, Athènes, Grèce, 92, 1988.

Suanez, S. and Sabatier, F.: Eléments de réflexion pour une gestion plus cohérente d'un système anthropisé: exemple du littoral du delta du Rhône/Ideas on the more coherent management of an anthropised system: the example of the coasts of the Rhône delta, In Revue de géographie de Lyon, Vol. 74, no. 1, Géographie des littoraux: la nature et les hommes, 7–25, https://doi.org/10.3406/geoca.1999.4925, 1999.

Syvitski, P. M. and Kettner, A.: Sediment flux and the Anthropocene, Philos. T. R. Soc. A, 369, 957–975, https://doi.org/10.1098/rsta.2010.0329, 2011.

Proc. IAHS, 377, 83–89, 2018
https://doi.org/10.5194/piahs-377-83-2018
© Author(s) 2018. This work is distributed under
the Creative Commons Attribution 4.0 License.

Open Access PIAHS

Satellite images survey for the identification of the coastal sedimentary system changes and associated vulnerability along the western bay of the Gulf of Tunis (northern Africa)

Abderraouf Hzami[1], **Oula Amrouni**[2], **Gheorghe Romanescu**[3], **Cristian Constantin Stoleriu**[3], **Alin Mihu-Pintilie**[4], **and Abdeljaouad Saâdi**[1]

[1]University of Tunis El-Manar, Faculty of Science, Laboratory of Mineral Resource and Environment, Tunis, Tunisia
[2]Laboratory of Marine Environment, National Institute of Marine Science and Technology, Tunis, Tunisia
[3]University Alexandru Ioan Cuza Iasi, Faculty of Geography and Geology, Department of Geography, Iasi, Romania
[4]University Alexandru Ioan Cuza Iasi, Interdisciplinary Research Department – Field Science, Iasi, Romania

Correspondence: Abderraouf Hzami (abderraoufhzami@gmail.com)

Received: 13 June 2017 – Accepted: 30 July 2017 – Published: 16 April 2018

Abstract. The aim of this study consists in testing the effectiveness of satellite data in order to monitoring shoreline and sedimentary features changes, especially the rapidly changing of Gulf of Tunis coast. The study area is located in the Gulf of Tunis western bay (Southern Mediterranean Sea) which is characterized by sandy beaches of Ghar Melah and Raoued (Medjerda Delta area). The aerial photographs and satellite imageries were used for mapping the evolution of shoreline. Diachronic data (satellite imagery, aerial photography and topo graphic maps) were used to monitor and to quantify, the evolution of the coastal areas. These thematic data were digitally overlaid and vectorised for highlighting the shoreline changes between 1936 and 2016, in order to map the rate of erosion and accretion along the shoreline. Results show that the accretion and degradation are related to the Medjerda: change of outlet in 1973 and impoundment of the Sidi Salem dam in 1982. We found that the general trend of the coastal geomorphic processes can be monitored with satellite imageries (such as Sentinel A2, Spots 4 and 5), due to its repetitive coverage along the time and their high quality concerning the spectral contrast between land and sea areas. Improved satellite imageries with high resolution should be a valuable tool for complementing traditional methods for mapping and assessing the sedimentary structures (such as shoreline, delta, marine bars), and monitoring especially the lowlands coastal areas (slightly eroded).

1 Introduction

The general characteristics of coastal erosion worldwide are described in terms of geography by the types of erosion, the causes which starts the erosion processes, and the effects generated by erosion processes. A shoreline is defined as the linear interface between land and water areas (Dolan et al., 1980). Shoreline is an element with a high spatial variability which imposes a rapidly changing for coastal landforms (Mujabar and Chandrasekar, 2013). Hence the efficient methods are needed to handle the spatial and temporal variability of coastal shoreline using GIS techniques. Thieler et al. (2009) developed an extension for ArcGIS software which allows automatic measurements for shoreline changes. Many researches utilized remote sensing data to analyse coastal environments: Louati et al. (2014), Oyedotun (2014, 2017), Thinh and Hens (2017). Many studies in Tunisia have shown the effect of coastal degradation related to different natural and anthropogenic factors (Paskoff, 1988; Oueslati, 2004,

2010; Halouani et al., 2011, 2013; Saïdi et al., 2012, 2013; Louati and Zargoun, 2013; Louati et al., 2014).

2 Study area

Gulf of Tunis is located in NE of Tunisia and is bordered to the east by Mediterranean Sea, between 37°10′ N–10°16′ E (Cape Farina) and 37°55′ N–10°18′ E (Cape Gammarth). The western bay of the Gulf of Tunis regular coastline consists in a series of lagoons disposed from north to south: such as Ghar Melah, Kalaât Andalous and Ariana. The coastline has a length of 40 km and is characterized by landforms such as coastlines, sandy features, sandy spit, river mouth deposit and dunes covered by forests.

Tidal range at the study area is low with amplitude of approximately 35 cm (Oueslati, 1993). Mean amplitude of semi-diurnal micro-tidal activities in the Gulf of Tunis measures 12–30 cm (El Arrim, 1996; Saïdi et al., 2012). The Western bay coastal system includes three main sedimentary systems with followed entities: Lagoons of Kalaât Andalous, delta of Medjerda and the coastal foredune of Raoued beach (Fig. 1). The hydrologic regime of the coastline is controlled by the Medjerda river which represents the most important river in Tunisia. During the last century, in the study area, the relative sea level has reached 11.5 mm year^{-1}; part of this value, such as 1.5 mm year^{-1}, has been attributed to eustacy (Pirazzoli, 1986). The spatial variability of sedimentary activities for the coastal system has a significant decline due to erosion processes: in the Medjerda Gammaret area between years 1887–1974, the coastline was registered a retreat rate of 0.12 to 1.11 m year^{-1}, and a decline of 3.73 to 9 m year^{-1} between 1974 and 2000 (Saïdi et al., 2014).

The Medjerda river (western bay) and Méliane river (eastern bay) supplies the most of continental supply for Gulf of Tunis coast. In this case, the impacts of climate change on the sedimentary dynamics of the catchment areas are subject to a very active research for a decade and which emphasise the reduction of fluvial fluxes of the river to the coastal environment of Gulf of Tunis (Arnell, 1999a, b; Oueslati, 2004; Saïdi et al., 2014).

3 Material and methods

In this study, three cloudless satellite images were used (SPOT1, SPOT4, and Sentinel A2). The shooting time, satellite information, and image resolution of each image are summarized in Table 1. As same resolution data is not available for the desired period but the multi temporal capabilities allow tracking of changes over a long time (80 years).

Pretreatment of imageries consisted in radiometrically and geometrical correction in order to minimize weathering effects on radiometric values (Fig. 2). The satellite imageries (SPOT1, SPOT4, Sentine A2), topographic maps and aerial photographies used in this study were ortho-rectified. The datum of cartographic and satellite and aerial imageries is World Geodetic System (WGS84), and the projected system is Universal Transverse Mercator. Firstly, the atmospheric effect where corrected by providing the dark object subtraction model (Chavez Jr., 1996), to compensate differences in shooting conditions and to calibrate the sensors.

The photo interpretation methods were used in order to draw line features (in vector format) for delineating the coastlines for each year corresponding to products listed in Table 1. Synthetic maps were obtained using ArcGIS 10.2 software. The diachronic synthesis based on shapefiles that contains coastlines had needed to evaluate the coastal dynamics (Robin, 2002).

The all graphical data (vector format) were used to calculate the spatial shoreline evolution using statistics provided by Digital Shoreline Analysis System (DSAS) (Thieler et al., 2009) in order to highlighting the differences between

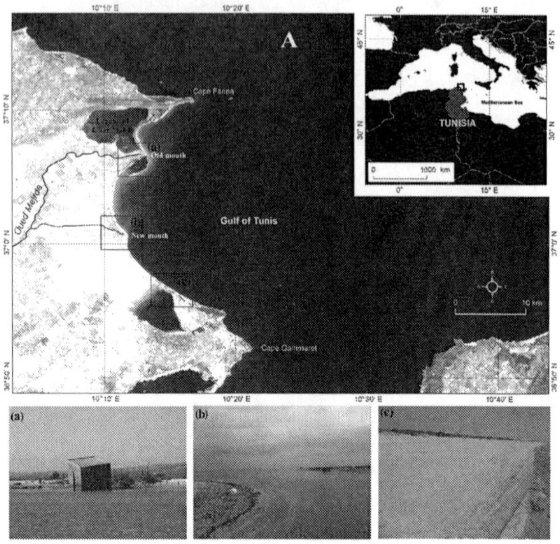

Figure 1. **(A)** Location map of the study site: the bay of Gulf of Tunis (Bay western). **(a)** Spit of Kalaât Andalous, **(b)** New mouth of Medjerda river and **(c)** Raoued beach.

Table 1. Satellite imagery and maps used in the study.

Spatial database	Date (year)	Resolution/ accuracy
Satellite imagery	Spot1 (04/09/1988)	20 m
	Spot4 (12/09/1999)	20 m
	SentinelA2 (16/04/2016)	10 m
Aerial photography	1974 (OTC) 2000 (CNCT)	2 m
Topographic maps	1974 (OTC) 1936 (OTC)	2 m

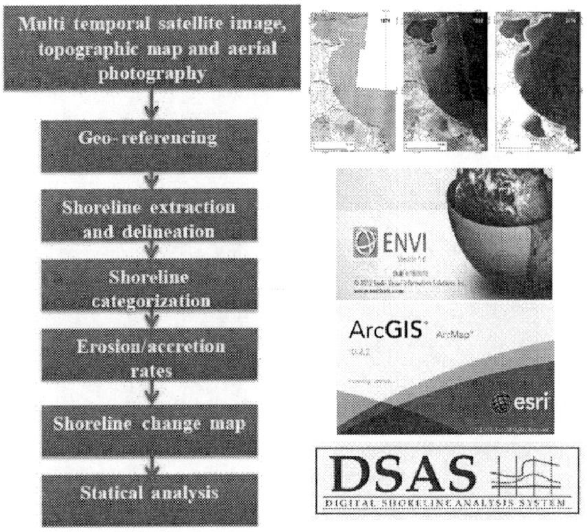

Figure 2. The workflow of methodology.

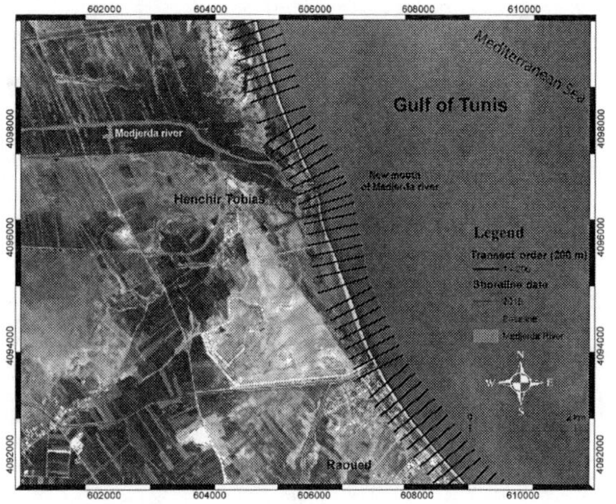

Figure 3. Shoreline extraction with transect and baseline using satellite imagery and maps in studied area.

coastline's positions. DSAS tool was used to create transversal transects over coastlines needed to analyse and calculate, in different points, the change rate at the specified time interval (Thieler and Danforth, 1994). For each transect, DSAS provides a calculation of erosion and accretion of shoreline (Bush and Young, 2009). Transects lines were generated automatically using DSAS with followed characteristics: lines with 1 km length placed at 200 m distance between them to study the changes that occurred along the delta of the Medjerda, and the western bay of the Gulf of Tunis (Fig. 3). Baselines can be placed ofshore or onshore of shorelines limit, however a baseline cannot be placed between shorelines. However in our study, the baseline was built offshore and parallel to the general trend of the coastline of gulf of Tunis. Transects will be cast perpendicular to this baseline (200 m spacing) and intersect the shorelines to establish measurement points (Himmelstoss, 2009).

Afterwards, all the shorelines specifically for each studied year were overlaid and spatially compared. During this stage can be emphasis the evolution of the coastline in the studied period. The quantitative and qualitative analysis is based on the statistics methods provided by DSAS, such as Net Shoreline Movement (NSM), End Point Rate (EPR) and Linear Regression (LRR).

The spatial error specific to georeferencing process for different products listed in Table 1 is estimated by the root mean square (RMS) and the maximum value is 2.33 m (Table 2). To calculate the margin of error of our photo-interpretation is based on the report from the USGS (2006) which expressed the calculation of the margin of error Eqs. (2), (3) and (4). However, we will not take into account the error of digitization for the development of the margin of error. In our graphics, we use a margin of error of $\pm 0.15\,\text{m}\,\text{year}^{-1}$. We take into account this type of error for a better comparison.

Table 2. Margin of error of photo-interpretation.

	Margin of error in meter				
Years	1936	1974	1988	1999	2016
RMS error			< 0.5		
Pixel error	1.8	1.62		1.42	1.21
Total error	2.3	2.12		1.92	1.71
Annual error ($\text{m}\,\text{year}^{-1}$) EPR: E			0.2		
Annual error ($\text{m}\,\text{year}^{-1}$) LRR: U			0.12		
Equation (1)	$E_{\text{sp}} = \sqrt{E_{\text{g}}^2 + E_{\text{d}}^2 + E_{\text{t}}^2 + E_{\text{p}}^2}$				
Equation (2)	$E_{\text{a}} = \frac{\sqrt{E_{\text{sp1}}^2 + E_{\text{sp2}}^2}}{\text{time}}$				
Equation (3)	$U = \frac{\sqrt{\sum_{i=1}^{n} C_i^1}}{n} + \frac{\sum_{i=1}^{n} B_i}{n}$				

4 Results and discussion

The temporal data comparison concerning the evolution of shoreline's positions using the statistic's variables, such as NSM and EPR, shows the importance of taking comparison of the temporal series of the evolution of the shoreline posi-

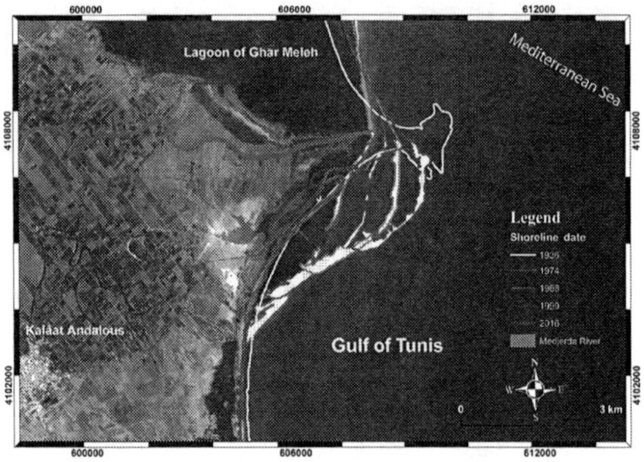

Figure 4. Multitemporal shoreline evolution of Kalaât Andalous.

tion using NSM and EPR statistical technics shows the importance of the different periods in this study.

Based on results it can be emphasis in the study area the presence of three dynamic zones:

1. Kalaât Andalous spit;
2. New mouth of Medjerda river;
3. Raoued beach.

The analyzing of the shoreline changes during 1936 to 2016 can be highlighted that erosion is significant, especially at the Kalaât Andalous spit with a severe erosion of up to $-25 \pm 0.15\,\mathrm{m\,year^{-1}}$ (-2 km) (Figs. 4, 5 and 6). Similar observation is reported along this zone coast by Saïdi et al. (2014) and Louati et al. (2014) with maximum rates of retreat respectively -17 ± 2.4 and $-20.7 \pm 3\,\mathrm{m\,year^{-1}}$. The decreasing of sediment flow at the old mouth of Medjerda river is generated by deviation of the new channel by hydrotechnical works in 1973. These hydrotechincal works stimulates the waves to erode and to carry out offshore the fluvial sediments from the old delta. The sediments under the effect of the coastal drift disposed on SE/NW direction facilitates the construction of barrier bar i.e. the sand spit of Kalaât Andalous.

During this period (1936–2016), the accretion is always occurred in the new mouth with a maximum rate of 320 m ($+4 \pm 0.15\,\mathrm{m\,year^{-1}}$) (Figs. 7, 8 and 9). During the floods of March 1973, Medjerda river had resorted to a change of bed choosing the channel of Henchir Tobias as an outlet.

The river's spillway is no longer powered by the alluvial contribution of Medjerda river. The results are shrinking beaches and coastal erosion. This phenomenon is evident in the aerial photographs of 1962 (Paskoff, 1985). On the other hand, the rate of accretion of the littoral has been eroded, at the mouth with an average of $-1.64 \pm 0.15\,\mathrm{m\,year^{-1}}$. This rate of erosion located at the river's mouth is explained by building of dam construction (El Aroussia in 1957, Bou

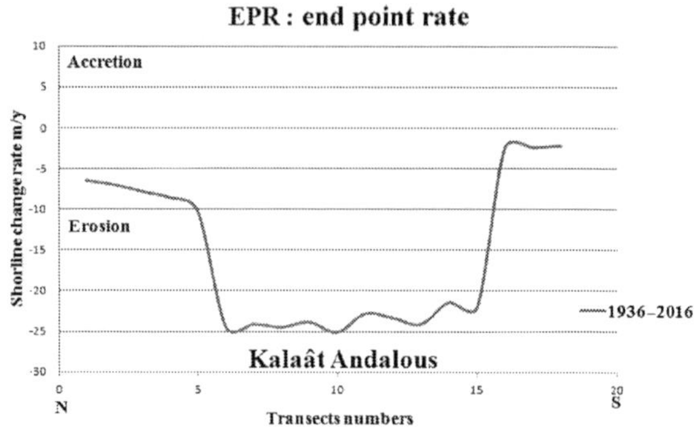

Figure 5. The rates of shoreline changes (EPR) in Kalaât Andalous.

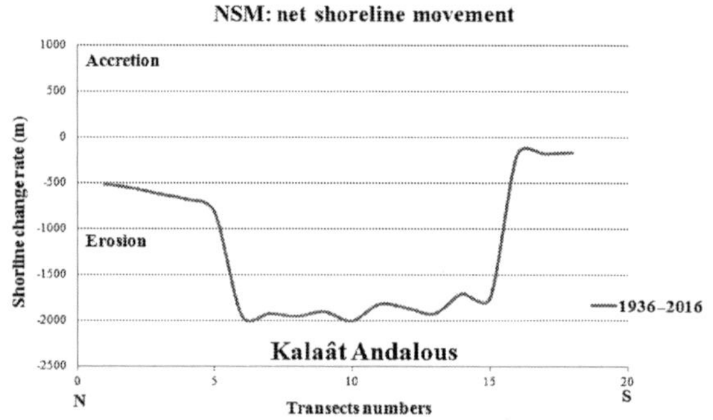

Figure 6. The rates of shoreline changes (MSN) in Kalaât Andalous.

Heurtma in 1976 and Sidi Salem in 1982) across the new Medjerda river restricts the flow of terrigenous sediments at the level of new mouth of Medjerda river. The Sidi Salem is characterized by a volume capacity of 555 million $\mathrm{m^3}$, a surface of 4.3 ha, also being the largest dam in the Tunis. Dam retain the terrigenous sediment discharge to the sea.

Between 1974 and 1988, the speed of advance of the shoreline continued during this period, especially in the mouth area with a maximum of 290 m. El Arrim (1996) reveals that the speed of the shoreline continued at a slower pace which gains of almost 270 m between 1977 and 1987 (El Arrim, 1996 in Oueslati, 2004).

Raoued beach, situated at the south east of the Medjerda delta, is an exceptional case in terms of fragility and management. The status of this area is favored, especially in its western part, by a large alluvial contribution occured by the new mouth of Medjerda river supply.

In this sector, we observe that the NSM is different for the two periods (1936–1974 and 1974–2016). Before 1974, this is a zone of erosion range with an average rate of -300 m ($-8 \pm 0.15\,\mathrm{m\,year^{-1}}$) (Figs. 10, 11 and 12). This erosion is due both to the effect of the longshore coastal drift from the SE to NW direction and this zone is highly erosional because of its exposure to the strong waves, winds and rip currents of

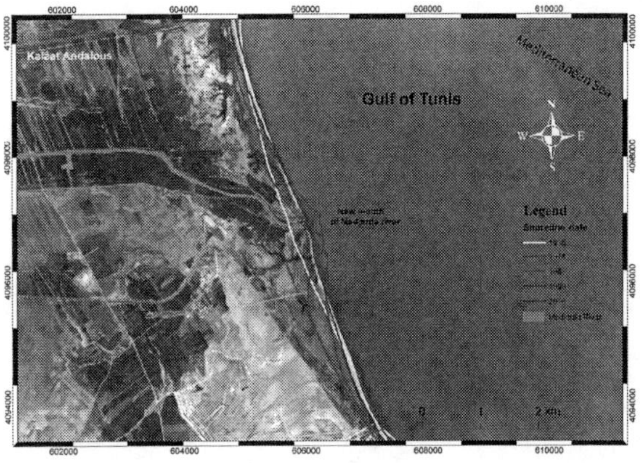

Figure 7. Multitemporal shoreline evolution of new mouth of Medjerda river.

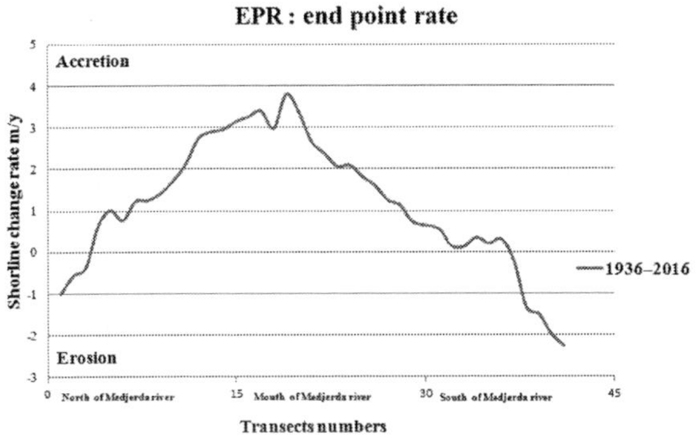

Figure 8. The rates of shoreline changes (EPR) in new mouth of Medjerda river.

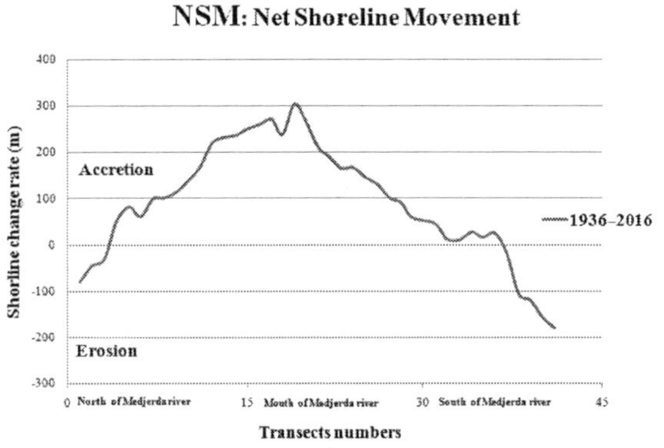

Figure 9. The rates of shoreline changes (MSN) in new mouth of Medjerda river.

Figure 10. Multitemporal shoreline evolution of Raoued.

the E to SE in summer, N to NE in winter. After 1974, this sector tends to be deposited, the values of erosion decreased significantly on all zone of Raoued ($-2.3 \pm 0.15\,\mathrm{m\,year^{-1}}$). This decrease in erosion rates is due to the deviation of Medjerda river in 1973. The natural river of Medjerda was abandoned, and the entire terrigenous flow now passes through an artificial canal of Henchir Tobias (2 km northern ward Raoued beach).

The main causes of erosion during this period are both related to natural and/or anthropogenic factors, especially by significant reduction of sediment supply caused by construction of numerous dams located in the catchment area of Medjerda river.

5 Validation of Digital Shoreline Analysis System (DSAS)

The LRR is determined by adjusting a least squares regression line at all the coastline points for a particular transect (Thieler et al., 2009). In the transect 116, the linear regression is the slope of the line with $y = 7.720x + 15\,953\,\mathrm{m\,year^{-1}}$. This calculation provides the standard error of the slope with the confidence interval of 95 %. Confidence in the analytical results is validated also by comparing this study with other researchers (Oueslati, 2004; Saïdi et al., 2014 and Louati et al., 2014), which indicated a good correlation. In the present study, spatial resolution (20, 10 and 2 m) is acceptable compared to previous studies (Louati and Zargouni, 2009; Louati et al., 2014) in this area using Landsat scenes with a spatial resolution of 30 m, which this increases the error in coastal areas.

6 Conclusion

The assessment of erosion and accretion processes through the transect lines using DSAS tool, applied on western bay of the Gulf of Tunis, provide valuable statistics upon coastline dynamics in terms of positional changes and in identification of depositional and denudational areas. Based on this study

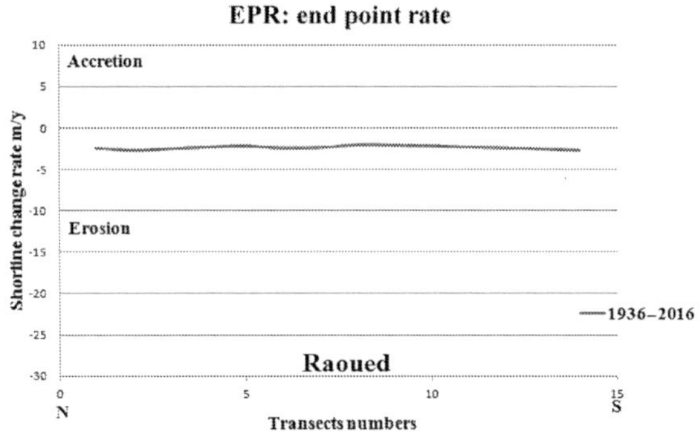

Figure 11. The rates of shoreline changes (EPR) in Raoued.

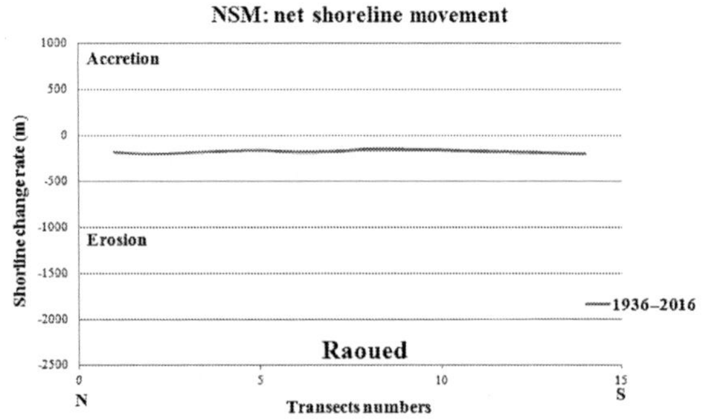

Figure 12. The rates of shoreline changes (MSN) in Raoued.

it can be concluded that DSAS will be useful for long-term (1936–2016) qualitative monitoring of shoreline evolution pattern in case lack of field data sources. In the studied period most of the beach underwent erosion ($-25 \pm 0.15\,\mathrm{m\,year^{-1}}$ in Kalaât Andalous sandy spits) while some part of the beach follow accretion trend ($+4 \pm 0.15\,\mathrm{m\,year^{-1}}$ in new mouth of Medjerda river). The variation of the morphology in the recent Medjerda river mouth was significant as well. The observed patterns of erosion and accretion along the bay shorelines resulted from both natural and human impacts strongly managed by human activities start to be more sensitive and vulnerable to natural erosion processus. Most of the shoreline was exposed to natural erosion processes induced by waves, tides and periodic storm surge. This study extends the analysis period by three years compared to the study conducted by Louati et al. (2014).

Data availability. Spot image: two images were provided by the laboratory of hydroscience Montpellier, in the project RYSCMED and were used in this work after the acceptance of the "ISIS" file.

Sentinel image: the Copernicus Open Access Hub (previously known as Sentinels Scientific Data Hub) provides complete, free and open access to Sentinel A2 (https://scihub.copernicus.eu/).

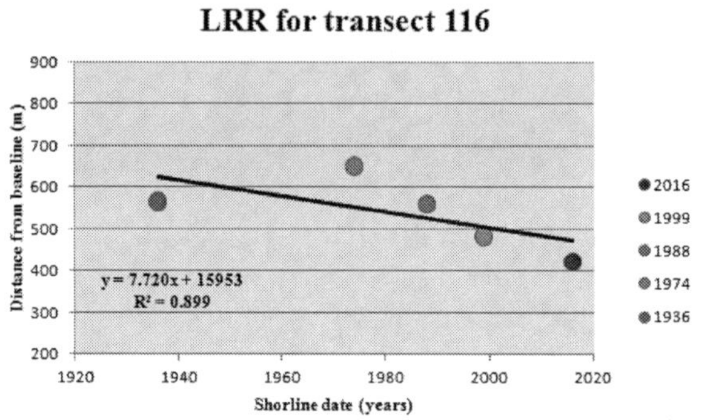

Figure 13. Linear regression of transect 116.

Competing interests. The authors declare that they have no conflict of interest.

Special issue statement. This article is part of the special issue "Water quality and sediment transport issues in surface water". It is a result of the IAHS Scientific Assembly 2017, Port Elizabeth, South Africa, 10–14 July 2017.

Edited by: Gil Mahe
Reviewed by: two anonymous referees

References

Arnell, N. W.: A simple water balance model for the simulation of streamflow over a large geographic domain, J. Hydrol., 217, 314–335, 1999a.

Arnell, N. W.: The effect of climate change on hydrological regimes in Europe: a continental perspective, Global Environ. Chang., 9, 5–23, 1999b.

Bush, D. M. and Young, R.: Coastal Features and Processes, in: Geological Monitoring, edityed by: Young, R. and Norby, L., Geological Society of America, Colorado, 47–67, 2009.

Chavez Jr., P. S.: Image-based atmospheric corrections. Revisited and improved, Photogramm. Eng. Rem. S., 62, 1025–1036, 1996.

Dolan, R., Hayden, B., May, P., and May, S.: The reliability of shoreline change measurements from aerial photographs, Shore and Beach, 48, 22–29, 1980.

El Arrim, A.: Etude d'impact de la dynamique sédimentaire et des aménagements sur la stabilité du littoral du golfe de Tunis, Thèse de Doctorat de Spécialité, Géologie, Faculté des Sciences de Tunis, p. 208, 1996.

Halouani, N., Fathallah, S., and Gueddari, M.: Beach and nearshore morphodynamic changes of the Tabarka coast, Northwest of Tunisia, Environmental Earth Sciences, 66, 1059–1069, 2011.

Halouani, N., Gueddari, M., and Frihy, O.: The Northwestern Mediterranean coast of Tunisia: wave processes, shoreline stability and management implications, Arabian Journal of Geosciences, 38, 1851–1860, 2013.

Himmelstoss, E. A.: DSAS 4.0 Installation Instructions and User Guide, in: 2009 Digital Shoreline Analysis System (DSAS) version 4.0 – An ArcGIS extension for calculating shoreline changes, edited by: Thieler, E. R., Himmelstoss, E. A., Zichichi, J. L., and Ergul, A., U.S. Geological Survey Open-File Report 2008-1278, 2009.

Louati, M. and Zargouni, F. : Modélisation topo-bathymétrique et transit sédimentaire. Exemple des plages sableuses de la baie de Tunis, Nord-Est de la Tunisie, Géomorphologie: relief, processus, environnement 3, 211–222, 2009.

Louati, M. and Zargoun, F.: Le littoral entre l'actuelle embouchure de l'oued Milianeet Soliman, Tunisie. Analyse de l'évolution du trait de côte par photo-interprétation et système d'information géographique, Géomorphologie: relief, processus, environnement, 19, 209–224, 2013.

Louati, M., Saïdi, H., and Zargouni, F.: Shoreline change assessment using remote sensing and GIS techniques: a case study of the Medjerda delta coast, Tunisia, Arabian Journal of Geosciences, 8, 4239–4255, https://doi.org/10.1007/s12517-014-1472-1, 2014.

Mujabar, P. S. and Chandrasekar, N.: Shoreline change analysis along the coast between Kanyakumari and Tuticorin of India using remote sensing and GIS, Arabian Journal of Geosciences, 6, 647–664, 2013.

Oueslati, A.: Les côtes de la Tunisie: géomorphologie et environnement et aptitudes à l'aménagement, Série 2, Publication de la Faculté des Sciences humaines et sociales de Tunis, Géographie, Université de Tunis I., 1993.

Oueslati, A.: Littoral et aménagement en Tunisie Orbis, Tunis, 534 pp., 2004.

Oueslati, A.: Plages et urbanisation en Tunisie: des avatars de l'expérience du xxe siècle aux incertitudes de l'avenir, Méditerranée, https://doi.org/10.4000/mediterranee, 115, 103–116, 2010.

Oyedotun, T. D. T.: Shoreline geometry: DSAS as a tool for historical trend analysis, in: Geomorphological Techniques, British Society for Geomorphology, UK, 12 pp., avalable at: http://geomorphology.org.uk/sites/default/files/geom_tech_chapters/3.2.2_ShorelineGeometry.pdf, last access: 11 November 2014.

Oyedotun, T. D. T.: Historical Shoreline Changes as Indication of Geomorphic Phases in St Ives and Padstow Bays of Southwest England, Environmental Processes, 4, 273–282, https://doi.org/10.1007/s40710-017-0213-3, 2017.

Pirazzoli, P. A.: Secular trends of relative sea level change indicated by tide gauge record, J. Coastal Res., S1, 1–26, 1986.

Paskoff, R.: Géographie de l'environnement: problèmes d'utilisation des ressources et d'adaptation aux contraintes des milieux naturels: exemples tunisiens, Université de Tunis, 227 pp., 1985.

Paskoff, R.: Tunisia, in: Artificial structures and shorelines, 1st Edn., edited by: Walker, H. J., Springer, Dordrecht, The GeoJournal Library, 10, 269–271, 1988.

Robin, M.: Remote sensing, from satellites to GIS. A comprehensive analysis of the process of creating an essential type of geographic information, Nathan University, 318 pp., 2002.

Saïdi, H., Souissi, R., and Zargouni, F.: Impact of shore-parallel breakwaters on the micotidal coast of Hammam-Lif (North-East of Tunisia), Arabian Journal of Geosciences, 5, 345–352, 2012.

Saïdi, H., Souissi, R., and Zargouni, F.: Grain size characteristics of superficial sediments of the Gulf of Tunis (NE Tunisia), Arabian Journal of Geosciences, 7, 3365–3387, https://doi.org/10.1007/s12517-013-1008-0, 2013.

Saïdi, H., Souissi, R., Louati, M., and Zargouni, F.: Morphologic changes and sedimentary budgets along a Mediterranean coastline with a sand spit: case of the littoral fringe Sidi Ali El Mekki-Gammarth (NE Tunisia), Rend. Fis. Acc. Lincei, 25, 393, https://doi.org/10.1007/s12210-014-0314-0, 2014.

Thieler, E. R. and Danforth, W. W.: Historical shoreline mapping (II): application of the Digital Shoreline Mapping and Analysis Systems (DSMS/DSAS) to shoreline change mapping in Puerto Rico, J. Coastal Res., 10, 600–620, 1994.

Thieler, E. R., Himmelstoss, E. A., Zichichi, J. L., and Ergul, A.: Digital Shoreline Analysis System (DSAS) version 4.0 – An ArcGIS extension for calculating shoreline change: U.S. Geological Survey Open-File Report 2008-1278, current version 4.3, available at: https://woodshole.er.usgs.gov/project-pages/DSAS/version4/ (last access: September 2017), 79 pp., 2009.

Thinh, N. and Hens, L.: A Digital Shoreline Analysis System (DSAS) applied on mangrove shoreline changes along the Giao Thuy Coastal area (Nam Dinh, Vietnam) during 2005–2014, Vietnam Journal of Earth Sciences, 39, 87–96, https://doi.org/10.15625/0866-7187/39/1/9231, 2017.

Proc. IAHS, 377, 91–96, 2018
https://doi.org/10.5194/piahs-377-91-2018
© Author(s) 2018. This work is distributed under
the Creative Commons Attribution 4.0 License.

Sedimentary evolution and ecosystem change in Ahémé lake, south-west Benin

Ernest Amoussou[1,2], Henri S. Totin Vodounon[1,2], Expédit W. Vissin[2], Gil Mahé[4], and Marc Lucien Oyédé[3]

[1]Department of Geography and Land Use Planning, University of Parakou, BP 123 Parakou, Benin
[2]Laboratory Pierre PAGNEY, Climate, Water, Ecosystems and Development (LACEEDE), University of Abomey-Calavi, 03 BP1122 Cotonou, Benin
[3]Department of Earth Sciences, Faculty of Technical Sciences University of Abomey-Calavi, Benin
[4]IRD, Laboratory HydroSciences of Montpellier, University of Montpellier 2, Case courrier MSE, Place Eugène Bataillon, 34095 Montpellier CEDEX 5, France

Correspondence: Ernest Amoussou (ernestamoussou@gmail.com)

Received: 14 June 2017 – Revised: 14 February 2018 – Accepted: 19 February 2018 – Published: 16 April 2018

Abstract. Tropical moist ecosystems, such as Ahémé lake, south-west Benin, are increasingly marked by water degradation, linked with the activities of increasing riparian populations. The objective of this study is to analyze sedimentary dynamics and its influence on the changing ecosystem of Ahémé lake from 1961–2010. Data used to carry out the study are records of precipitation, flows, turbidity, suspended sediment, mineral elements and bathymetry. Grain size data from the sieving of sediment samples were used to interpret suspended solids distribution in the lake. Linear correlation coefficients were used to assess the degree of dependence between rainfall and runoff inputs to the lake. Lake depth measurements in some areas of the lake serve to determine the rate of infilling. The sorting index was used to highlight the distribution and origin of sediments in the lake. The results show a degradation of the lake Ahémé ecosystem characterized by infilling of its bed, a high correlation ($r = 0.90$) between rainfall and runoff, seasonal change in physicochemical parameters (total suspended sediment decrease by −91 %) and decrease in fish production by $135.8\,\mathrm{t\,yr^{-1}}$. The highest mean suspended sediment concentrations in lake inputs occur during high water periods ($123\,\mathrm{mg\,L^{-1}}$) compared to low water periods ($11.2\,\mathrm{mg\,L^{-1}}$).

1 Introduction

Ahémé lake, in Benin western Africa, and its biodiversity attract enormous interest from people, local residents and scientists in particular (Amoussou et al., 2016). The riparian population interest is in the supply of fish resources but the ecosystem of Ahémé lake is in constant degradation. The lake dynamics and its water resources are linked to climatic and hydrological variability (Amoussou et al., 2007). Analysis of the dynamics of rivers and water bodies (Vissin, 1998) is necessary to meet the main objectives of the Global Energy and Water Experiment (GEWEX) and the Tropical Atmosphere and Hydrologic Cycle (CATCH) projects.

Eutrophication of Ahémé lake (Oyédé, 1991; Amoussou, 2004) leads to siltation, excessive concentration of chemical elements, resulting in sometimes a significant presence of algae or aquatic plants, and also absence of certain plant species along the lake shores inducing hydraulic erosion.

These combined pressure factors could lead to destruction of the ecological habitats of fish species and consequently disrupt socio-economic activities. Commercial fishing is declining as a result of lower fisheries yields. Thus, the balance between natural resources and human population demands or needs is being compromised. This study aims to analyze the evolution of sedimentary and environmental parameters from 1961 to 2010 and their impacts on the ecosystem of Ahémé lake. This work brings together hydrometric, sedimentary and water quality information to help understand ecosystem changes and its impacts in Ahémé lake.

Published by Copernicus Publications on behalf of the International Association of Hydrological Sciences.

2 Data and methods

2.1 Study site

Ahémé lake is located between 6°20′ and 6°40′ N and 1°55′ and 2° E (Fig. 1a).

Ahémé lake is located in a depression between the Allada and Comè plateaux (Fig. 1b). The Couffo River flows into Ahémé lake from the north. Ahémé lake exchanges water in the south with the Mono and Sazoué rivers which are connected to a coastal lagoon and the Atlantic Ocean through the 24 km long Aho Channel. During the wet season freshwater water from Ahémé lake flows southwards in the channel to the lagoon and Atlantic Ocean. However, during the dry season the flow in the channel reverses, resulting in increased salinity in the southern part of the lake. The width of the lake is ∼ 3.4 km at the latitude of Guézin. The surface area of Ahémé lake is between 70 and 100 km^2 in the dry and rainy seasons, respectively (Le Barbé et al., 1993).

Ahémé lake is influenced by a rather dynamic lagoon system that favors variations in pH, temperature and salinity due to its opening on the Atlantic Ocean. The contribution of Mono River to the lake is most significant during floods or periods of high water levels (Pliya, 1980; Oyédé, 1983; Amoussou et al., 2007).

2.2 Data

Rainfall data from stations at Athiémé, Grand Popo, Bopa, Allada and Ouidah (see Fig. 1a) over the period 1961–2010 were extracted from the database of METEO BENIN. Flow data records of the Couffo River at Lanta and Mono River at Athiémé over the period 1961–2005 (data are not available for 2006–2010) were collected by the Hydrology Department of the Directorate General for Water. Information on sediment dynamics in Ahémé lake was available as: (1) sediment depth data collected only in 1991 and 1999 were extracted from Oyédé et al. (2007); (2) annual suspended solids concentrations measured in 1999 and 2007 at Guézin (Roche International, 1999; Amoussou et al., 2007). These were complemented by seasonal (rainy and dry) bathymetry measurements in 2003 and 2006 in Ahémé lake. Physico-chemical measurements were made at localities on the east and west shores of the lake in: (1) October 2000 – pH only; and (2) September 2002 – salinity and pH measured using a WTW 340i handheld pH/conductivity meter. Sediment samples collected during hand dredging were analyzed by sieving using the AFNOR (French Standardization Association) method (AFNOR, 1996). Sediment weighing was carried out on a Shimadzu BX3200D (dual range: 3200/600 g and resolution: 0.1/0.01 g).

2.3 Methods

Linear correlation coefficient were calculated between mean monthly rainfall at Athiémé, Grand Popo, Bopa, Allada and

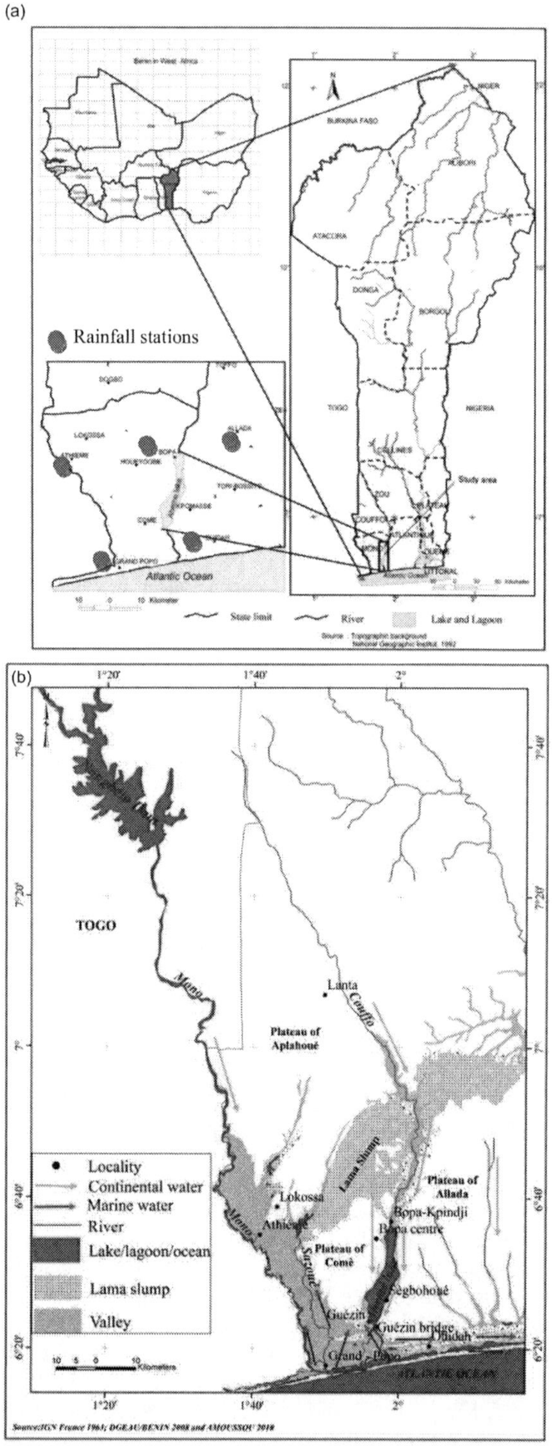

Figure 1. **(a)** Location of Ahémé lake in south-west of Benin. **(b)** The flow direction of the Mono-Couffo rivers and Ahémé lake.

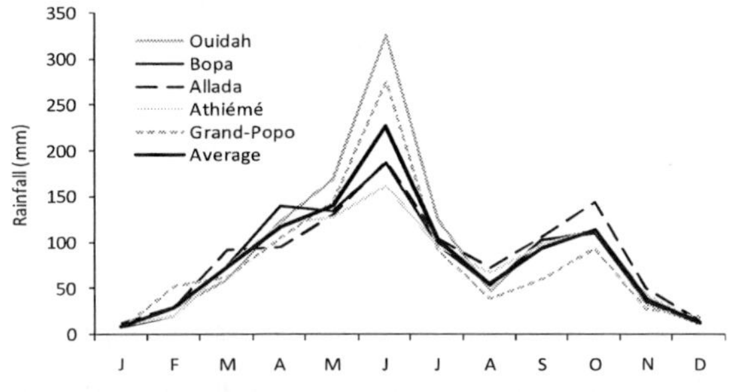

Figure 2. Mean monthly rainfall at the stations around Ahémé lake over the period 1961–2010.

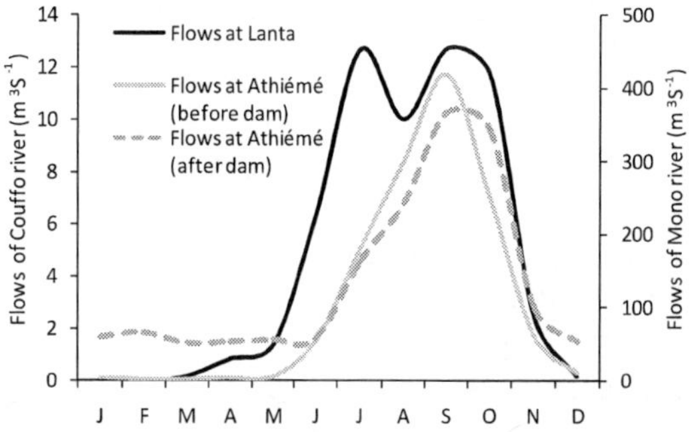

Figure 3. Mean monthly flow of the Mono River at Athiémé and Couffo River at Lanta linking Ahémé lake over the period 1961–2005.

Ouidah station and mean monthly flow at stations Athiémé and Lanta for the period 1961-2005 (see Figs. 2 and 3). Bravais-Pearson's correlation coefficient is calculated with 0.05 significance level.

Assessment of the sedimentary evolution of Ahémé lake was conducted to determine whether sediment depositions or erosion was occurring or whether there was sedimentary balance. The lake cross-sectional area is calculated from the depth and width of the wetted section. Thus, the decrease or increase in section area is related to depth, because from one year to the next in the same season, the width varies very little.

The bathymetry is measured by boat over 2.5 km at Bopa Kpindji, Bopa Centre and Sègbohouè (Fig. 1b), from the east to west bank. On this cross section, the measurement depths of Ahémé lake is doing each 2 m distance with graduated wooden ruler on dry and rainy seasons.

These measurements referred to a standard water level (0.5 m) at the gauging station under Guézin bridge (Fig. 1b) and formed the basis for estimating the sediment depth data shown in Table 1. The Sorting index (S_0) was used to determine the distribution and origin of sediments in the lake through the formula:

$$S_0 = \frac{q_3}{q_1},$$

where q_1 and q_3 are the first and third quartile, respectively, of the grain size distribution.

- If $S_0 = 1$ or close to 1: the sediment is homogeneous (well sorted), corresponding to a steep gradient of the grain size curve (Ben Amor et al., 2003; Marc and Emblanch, 2005);

- If $S_0 <$ or > 1: the sediment is poorly sorted out, corresponding to a low gradient of the grain size curve (Ben Amor et al., 2003; Marc and Emblach, 2005).

3 Results and discussion

3.1 Rainfall and river flow variation

Figures 2 and 3 show the mean monthly rainfall and flows of rivers feeding the Ahémé ecosystem. The annual rainfall regime is bimodal, characterized by an important supply of water over May–July (rainy season) and September–October (small rainy season).

This seasonal climate variability results in the transport of dried sediment from the banks into Ahémé lake. Sediment and water inflow from the Couffo River to Ahémé lake is less important than that of the Mono River which has a higher flow (annual mean flow is 5.1 $m^3 s^{-1}$ at Lanta and 114.4 $m^3 s^{-1}$ at Athiémé over 1961–2005).

Correspondence of flows with the rainy season confirms the high rainfall vs. flow Bravais-Pearson's correlation coefficient ($r = 0.90$) in the study area.

Moreover, the lake receives a large volume of water during September–October when the rivers entering the lake reach their maximum annual flow and also contain their highest suspended solids concentration as reported by Amoussou (2004). The Couffo river has much greater seasonal variability in flow compared to the Mono river especially since construction of the dam on the Mono river at Nangbéto (7°25′25.40″ N; 1°26′5.82″ E) in September 1987. The difference in the hydrological flow regimes in the Couffo and Mono rivers and the effect of the Nangbéto dam on dry season flow are evident in Fig. 3. Base flow in December to April at Athiémé was 4.22 $m^3 s^{-1}$ before the dam construction and increased to 57.26 $m^3 s^{-1}$ after the dam construction. The dam has had a major impact on water and sediment flows from the Mono river to the Ahémé lake ecosystem, including increased bank erosion downstream of the dam (Oyédé, 1991; Amoussou, 2010).

Table 1. Sedimentation evolution in some parts of the Ahémé lake from 1991 to 1999.

Site	Sectional areas of sediment (m^2)			
	1991	1999	Difference (1991–1999)	Variation rate (%)
Bopa Kpindji	2405	2499	+94	3.9
Bopa Centre	5164	4919	−245	−4.7
Sègbohouè	2899	2744	−155	−5.3

Source: Oyédé et al. (2007)

3.2 Morphodynamics of Ahémé lake

Data from the years 1991 and 1999 (Oyédé et al., 2007) revealed declining water depths in some areas of Ahémé lake. Comparison of the depths between 1991 and 1999 in a few areas of the lake, in the north (Bopa Kpindji and Bopa Centre) and the south (Sègbohouè), revealed infilling (Table 1) from the east bank to the west bank. In the north of the lake, there was erosion at Bopa Centre from 1991 to 1999, together with deposition at Bopa Kpindji, especially on the eastern bank.

Infilling of Ahémé lake, as indicated by some of the data presented here, is one of the factors of the lake impoverishment contributing to fish species migration and ecosystem degradation.

Suspended solid inflows to the lake are more important in the rainy season than in the dry season, as demonstrated by measurements carried out by Roche International (1999) and Amoussou et al. (2007), who showed that at Guézin bridge, the daily suspended solids concentration measured during the rainy season varied from 81 to 165 $mg L^{-1}$ (mean ~ 123 $mg L^{-1}$) while in the dry season they range from 3.5 to 19 $mg L^{-1}$ (mean ~ 11.2 $mg L^{-1}$).

3.3 Granulometric analysis

Figure 4 shows the parabolic curves of granulometric variation for sediment samples from the coastal lagoon and Ahémé lake, based on the sediment weight data transformed into a cumulative percentage.

The proportion of sands (> 0.5 mm diameter) is greater (80 %) in the coastal lagoon than in Ahémé lake (40 %). The particle size range is narrower in Ahémé lake and is dominated by fine particles, indicating the transport of particles in suspension. These results are consistent with those obtained by Yalin and Karahan (1979) on the secondary tributary of the Loire river and by Fournier (2004) on the Durance river, both in France, and by Degoutte (2006) on embankment dams. The textural heterogeneity of the sediments is also due to their origin: either from the crystalline basement geology or sedimentary rocks affected by erosion, or from the sea.

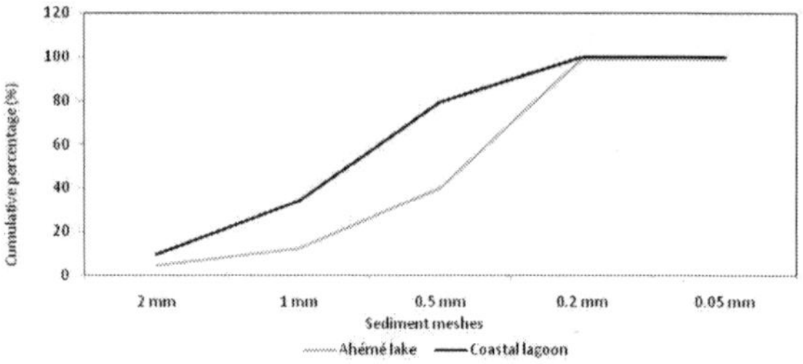

Figure 4. Sediment granulometric curves in the coastal lagoon and Ahémé lake (Source: Amoussou et al., 2016).

Values of the Sorting Index (S_0) were > 1, showing that deposited sediments are poorly sorted in both Ahémé lake ($S_0 = 2.15$) and in the lagoon ($S_0 = 1.71$). This can be explained by the effect of a range of processes, including low roughness of the floodplain, flocculation due to saline conditions at high tide, solid inputs generated by anthropogenic activities and the transport and deposition of the majority of sediment as bed load mobilised during the rainy season. These results are consistent with those of Cerdan et al. (2002) and Amoussou (2010), who reported that, because of the absence of vegetation, in the rainy season sediment particles are mobilized and deposited on the lake bed, modifying the lithofacies of the bottom. Large sediment particles are deposited on the river banks whilst finer sediment particles are transported by the river into Ahémé lake.

3.4 Evolution of physicochemical parameters

Salinity measurements at the entrance to the lake (Guézin first bridge) in the north (Bopa Agonsa) and on both banks of Ahémé lake in September 2002 (Fig. 5) show high salinity even during the rainy season. This is caused by the quasi-permanent opening of the Mono River mouth on the coastal lagoon. It could also be attributed to a decrease of rainfall around the lake in the years 2000 (−26 to −29 %), 2001 (−8 to −13 %) and 2002 (−10 to −15 %), compared with the annual mean rainfall for the period 1961–2010 (968.42 mm), providing less rainfall for dilution of salinity. Salinity values measured in the lake in September 2002 (at the beginning of the floods of the Mono and Couffo rivers, where salinity is almost zero) are much higher than those recorded in other parts of the drainage basin (Oyédé, 1981, 1983, 1991) and other rivers in Benin (Amoussou, 2003; Amoussou et al., 2007).

Increasing salinity can result in enhanced flocculation of terrigenous sediment in the lake and subsequent deposition within the lake, contributing to sediment infilling.

Analysis of the pH data (Fig. 6) allows assessment of the suitability of the lake water quality for aquatic ecology and fish species. The mean pH at the localities ranges from 7.2 to 8.2 in October 2000, indicating a basic environment dur-

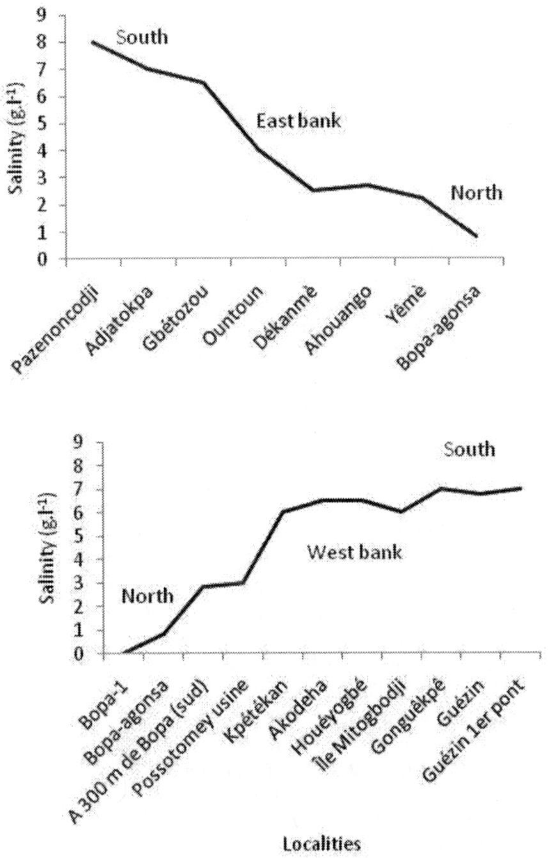

Figure 5. Spatial variation in salinity from South to North on the eastern and western bank of Ahémé lake in September 2002 (Source: Amoussou, 2003).

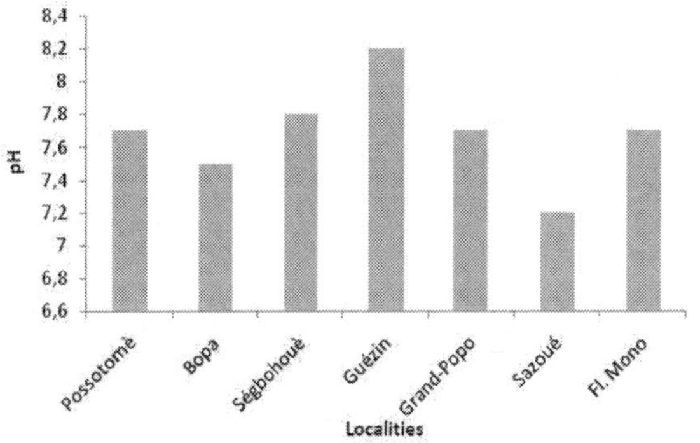

Figure 6. Spatial variation of pH measured in Ahémé lake in October 2000 (Source: PAZH, 2002).

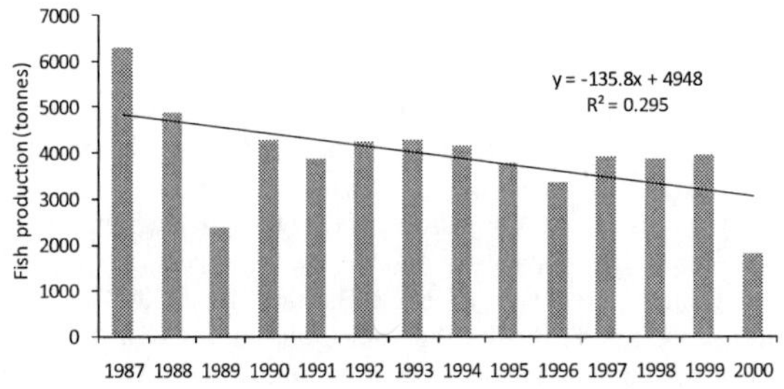

Figure 7. Annual fish production in Ahémé lake from 1987 to 2000 (Source: Fish farming Directorate, 2002).

ing high water periods. In the lake area, pH is quite variable, as also shown by Houadégla (1991) in Nokoue lake (6°25′23.71″ N; 2°26′26.30″ E) in south-east Benin, which is also in the subequatorial climate domain.

Lake pH values depend on the time and location of measurement, since they can be affected by solar insolation, intensity of chlorophyll assimilation, respiration of animals and metabolism of lower aquatic organisms. The high pH at Guézin (8.2) is attributed to the fact that saline marine waters necessarily pass through this point before reaching the lake.

Degradation of the Ahémé lake ecosystem due to eutrophication, low dissolved oxygen concentration (Dèdjiho et al., 2013; Dimon et al., 2014) and over-fishing resulting in lower fish reproduction rates (not demonstrated in this study) has affected fish production (Fig. 7) with a decline in annual fish production from 6298 t in 1987 to 1813 t in 2000.

4 Conclusion

Ahémé lake is characterized by a bimodal rainfall regime (two dry seasons and two alternating rainy seasons) and the unimodal flow regime of the Mono (with increased base flow as the result of dam construction) and Couffo Rivers. The high flows coincide with the heavy rains occurring in the lake's catchment. The strong flow / rain relationship reflects the dependence of the flow on rainfall and influences the variation of water physicochemical parameters. Infilling of the lake is confirmed by the continuing decrease in lake depth by sediment with a narrower range of particle size in the lake than in the coastal lagoon.

Data availability. The data are not publicly accessible:

- Climatology data are been propriety of Agence Météo Benin,
- Fish data are been propriety of Fish direction.

Competing interests. The authors declare that they have no conflict of interest.

Special issue statement. This article is part of the special issue "Water quality and sediment transport issues in surface water". It is a result of the IAHS Scientific Assembly 2017, Port Elizabeth, South Africa, 10–14 July 2017.

Acknowledgements. Authors wish to thank also the Agency of Météo Benin and the Fish directorate for providing data. The authors wish to thank the Associated Editor (Kate Heal), and an anonymous reviewer for their useful comments on the manuscript.

Edited by: Kate Heal
Reviewed by: two anonymous referees

References

AFNOR: NF P94-056. Sols: reconnaissance et essais. Analyse granulométrique. Méthode de tamissage à sec après lavage. Soil: investigation and testing. Granulometric analysis. Dry sieving method after washing, France, 15 p., 1996.

Amoussou E.: Dynamique hydro-sédimentaire et mutations des écosystèmes du lac Ahémé. Mémoire de maîtrise, DGAT/FLASH/UAC, 103 p., 2003.

Amoussou, E.: Systèmes traditionnels de gestion durable du lac au Bénin, in: Développement durable: leçons et perspectives, Acte de Colloque AUF Ouagadougou, Burkina Faso, 263–270, 2004.

Amoussou, E.: Variabilité pluviométrique et dynamique hydro-sédimentaire du bassin-versant du complexe fluvio-lagunaire Mono-Ahémé-Couffo (Afrique de l'Ouest), PhD thesis, Université de Bourgogne, Dijon, CRC – CNRS- UMR5210, 313 p., 2010.

Amoussou, E., Vissin, E. W., Houssou, C. S., and Boko, M.: Changes in Physicochemical parameters and productivity of Ahémé lake, in: 14th Stockholm Water Symposium: Drainage basin management-regional approaches for food and urban security, Abstract volume, 2004, Stockholm Water Symposium, 159–160, 2004.

Amoussou, E., Oyédé, L. M., and Boko, M.: Variabilité pluviométrique et flux de turbidité dans le complexe Chenal Ahôlac Ahémé au Bénin (Afrique de l'Ouest), Actes du XXème colloque AIC, in: Climat, Tourisme et Environnement de Carthage, Carthage, Tunisie, 81–86, 2007.

Amoussou, E., Totin, V. S. H., Tohozin, Y. A., Oyédé, M. L., and Boko, M.: Traditional adaptation strategies to hydrosystem degradation for sustainable management of the Ahémé lake in Benin (West Africa), European Scientific Journal of European Scientific Institue (ESI), 12, 352–365, 2016.

Ben Amor, R., Brahim, M., and Gueddari, M.: Essai d'interprètation de la dynamique sédimentaire par l'analyse granulométrique et minéralogique au large du Golfe de Gabès, Bull. Inst. Natn. Sci. Tech. Mer de Salammbô, 30, 143–151, 2003.

Cerdan, O., Le Bissonnais, Y., Souchere, V., Martin, P., and Lecomte, V.: Sediment concentration in interrill flow interactions between solid surface conditions, vegetation and rainfall, Earth Surf. Proc. Land., 27, 193–205, https://doi.org/10.1002/esp.314, 2002.

Dèdjiho, C. A., Mama, D., Tomètin, L., Nougbodé, I., Chouti W., Sohounhloué, C. K. D., and Boukari, M.: Évaluation de la qualité physico-chimique de tributaires d'eaux usées du lac Ahémé, Benin, J. Appl. Biosci., 70, 5608–5616, 2013.

Degoutte, G.: Diagnostic, aménagement et gestion des rivières: hydraulique et morphologie fluviales appliquées, Editions Tec & Doc, Lavoisier, France, 394 p., 2006.

Dimon, F., Dovonou, F., Adjahossou, N., Chouti, W., Mama, D., Alassane, A., and Boukari, M.: Caractérisation physico-chimique du lac Ahémé (Sud Bénin) et mise en relief de la pollution des sédiments par le plomb, le zinc et l'arsenic, J. Soc. Ouest-Afr. Chim., 37, 36–42, 2014.

Fish farming Directorate: Annual fish production in Ahémé lake from 1987 to 2000, cotonou, Bénin, 10 p., 2002.

Fournier, L.: Modélisation de la production des apports sédimentaires dans le bassin de la Durance, Mémoire de DEA, LNHE, CRD-EDF, Université Pierre et Marie Curie, Paris, France, 56 p., 2004.

Houadégla, W. A.: Rythmes climatiques et productions halieutiques au Bénin: cas du "lac" Nokoué, Mémoire de maîtrise en DGAT/FLASH/UNB, 126 p., 1991.

Le Barbé, L., Alé, G., Millet, B., Texier, H., Borel, Y., and Gualde, R.: Les ressources en eaux superficielles de la République du Bénin, Edition ORSTOM, France, 540 p., 1993.

Marc, V. and Emblanch, C.: Cours de méthode de calcul des paramètres granulométriques, France, 2005.

Oyédé, L. M.: Sédimentation margino-littoral au débouché du "lac" Ahémé (Bénin Afrique de l'Ouest), Mémoire de DEA, Science de la Terre, Dijon, 44 p., 1981.

Oyédé, L. M.: Un exemple de sédimentation biodétrique quaternaire dans le domaine margino – littoral en climat tropical humide: le "lac" Ahémé – (Bénin – Afrique de l'Ouest), PhD thesis, Université de Dijon, Paris, 171 p., 1983.

Oyédé, L. M.: Dynamique sédimentaire actuelle et messages enregistrés dans les séquences quartenaires et néogènes du domaine margino littoral du Bénin (l'Afrique de l'Ouest), PhD thesis, Université de Bourgogne, Paris, 302 p., 1991.

Oyédé, L. M., Kaki, C., and Laïbi, R.: Environnement sédimentaire, morphologie et faciès du lac Ahémé dans le complexe lagunaire sud-ouest béninois, Annales des Sciences agronomiques du Bénin, 9, 75–98, 2007.

PAZH: La pollution dans les zones humides du Sud-Bénin: état actuel, impacts, stratégies de suivi et de lutte. Rapport Intérimaire, 111, 42 p., 2002.

Pliya, J.: La pêche dans le Sud-Ouest du Bénin. Etude de géographie appliquée sur la partie continentale et maritime, AGECOOP, Paris, 293 p., 1980.

Roche International: Etude de projet d'aménagement des plans d'eau du Sud-Bénin, Volume III, Tome III, Rapport final, 101 p. + annexes, 1999.

Vissin, E. W.: Contribution à l'étude du fonctionnement hydrologique du bassin de la Sota, Mémoire de maîtrise, FLASH/UNB, 80 p. + annexes, 1998.

Yalin, M. S. and Karahan, E.: Inception of sediment Transport, J. Hydr. Eng. Div.-ASCE, 105, 1433–1443, 1979.